BEST MANAGEMENT PRACTICES
FOR VEGETATION MANAGEMENT

June 2005 (Revised December 2015)

Los Angeles County Weed Management Area

Carl Bell
Dean Lehman

Ellen Mackey, editor

Please cite as:

Bell, Carl and Dean Lehman. 2015. *Best Management Practices for Vegetation Management (revised).* Ellen Mackey, editor. Los Angeles County Weed Management Area. Los Angeles, California.

Cover photo:

Foeniculum vulgare (fennel), Kerstin Waurick, 2015.

Contents

CHAPTER 1. INTRODUCTION 1

CHAPTER 2. MANAGEMENT METHODS 7
    PREVENTION 8
        A. Sanitation 8
        B. Quarantine 10
        C. Education and Outreach 11
    ERADICATION 13
    CONTROL 14
        A. Physical Methods 15
            A.1. Hand Pulling and Hoeing 15
            A.2. Fire 16
            A.3. Flaming 17
            A.4. Steaming/application of hot water 18
            A.5. Foaming 19
            A.6. Tillage (cultivation) 20
            A.7. Mowing and shredding 21
            A.8. Mulches 23
            A.9. Soil Solarization 24
            A.10. Structural 25
            A.11. Weed Mats 26
        B. Cultural Methods 27
            B.1. Native Planting 27
            B.2. Living mulches, cover crops, and nurse crops 28
        C. Biological Methods 29
            C.1. Classical biological control 29
            C.2. Grazing and other forms of herbivory 30
        D. Chemical Methods 31
        E. Organic Methods 33
    RESTORATION 35

CHAPTER 3. COMPONENTS OF A VEGETATION MANAGEMENT PLAN 39
    PUBLIC ROADSIDES 39
    WILDLANDS 41

APPENDIX A. HOW PESTICIDES ARE REGULATED WITHIN CALIFORNIA 43

APPENDIX B. SUMMARY OF REGULATIONS AND RESOURCE AGENCIES 49

APPENDIX C. SUMMARY OF REGULATIONS FOR ORGANIC PESTICIDES 51

GLOSSARY 55

BIBLIOGRAPHY 59

INTERNET RESOURCES 63

Contributors

Carl Bell
Principal
Southern California Invasive Plant Management
www.socalinvasives.com

Mary Ann Bennett
Landscape Architect Associate
Architectural Engineering Division
Los Angeles County Department of Public Works
www.ladpw.org

Amy Bradsher
Marketing Director
Organic Materials Review Institute (OMRI)
www.omri.org

Jason Casanova
GIS Programs Manager
Council for Watershed Health
www.watershedhealth.org

Sabrina L. Drill, Ph.D.
Natural Resources Advisor
Los Angeles and Ventura Counties
University of California Cooperative Extension
http://celosangeles.ucanr.edu

Roger A. Haring
CCA/QAL/ Project Coordinator

Irina C. Irvine, Ph.D.
Restoration Ecologist
Santa Monica Mountains National Recreation Area
www.nps.gov/samo

Jim Hartman
Deputy Agricultural Commissioner/Sealer
Integrated Pest Management Division
Los Angeles County Agricultural Commissioner/Weights
& Measures
http://acwm.lacounty.gov

Danielle LeFer, Ph.D.
Conservation Director
Palos Verdes Peninsula Land Conservancy
www.pvplc.org

Dean Lehman
District Engineer
Los Angeles County Department of Public Works
www.ladpw.org

J. Lopez
Assistant Chief
County of Los Angeles Fire Department
Prevention Services Bureau
Forestry Division
Natural Resources Section
www.fire.lacounty.gov

Ellen Mackey
Senior Ecologist
Metropolitan Water District of Southern California
www.mwdh2o.com
emackey@mwdh2o.com

Bill Neill
Principal
Riparian Repairs

Kelly Schmoker, M.S.
Senior Environmental Scientist (Specialist)
Habitat Conservation Planning
California Department of Fish and Wildlife
www.wildlife.ca.gov

Raymond B. Smith
Bureau Chief
Weed Hazard and Integrated Pest Management Division
Los Angeles County Agricultural Commissioner/Weights
& Measures
http://acwm.lacounty.gov

Rachel Stoughton
Mountains Restoration Trust
www.mountainstrust.org

Richard Takata
Deputy Agricultural Commissioner/Sealer
Integrated Pest Management Division
Los Angeles County Agricultural Commissioner/Weights
& Measures
http://acwm.lacounty.gov

Rachel Wing
Park Ranger
Monrovia Hillside Wilderness Preserve
Department of Community Services, City of Monrovia
www.cityofmonrovia.org

Chapter 1. Introduction

Arundo donax (giant reed) control at Whittier Narrows Recreation Area. (Photo courtesy of Bill Neill, Riparian Repairs.)

The Los Angeles County Weed Management Area (LA WMA) supports weed and invasive plant control based upon Integrated Pest Management (IPM) strategies. Weed Management Areas (WMAs) are local organizations that bring together landowners and managers (private, city, county, state, and federal) in a county, multicounty or other geographic area to coordinate efforts and expertise against invasive weeds. Through this document, the LA WMA intends to inform individuals, businesses, government agencies and non-governmental organizations on all of the currently known methods of vegetation management, effectiveness of the technique, cost, safety, and potential environmental impacts for those methods suitable for LA County. **This document is designed to assist landowners and managers in using these Best Management Practices (BMPs) tailored to their specific needs for an integrated vegetation management approach.** To accomplish this we must explain three separate but related topics of integrated vegetation management: vegetation control methods, the suitable sites for these methods, and the pros and cons (risks/benefits) for each of these methods. Suitable sites include: wildlands, rights of way (roads, flood control, utility, easements), private property (small, large, individually owned, owned by conservancy or preserve), urban, rural, parklands/open space, riparian, aquatic, wetlands, public (federal, state, local, tribal, schools/ universities, water districts, special districts), and ornamental landscapes.

Definitions

According to the Weed Science Society of America, a weed (invasive plants are weeds of natural areas) is, "Any plant that is objectionable or interferes with the activities or welfare of man" (Vencill 2002). The Merriam-Webster Online Dictionary defines a weed as "a plant that is not valued where it is growing and is usually of vigorous growth, especially one that tends to overgrow or choke out more desirable plants". The Concise Oxford Dictionary states that a weed is "a wild plant growing where it is not wanted."

In Weed Ecology (Radosevich, Holt, and Ghersa 1997), vegetation management is a strategy that fosters beneficial vegetation along with suppressing undesirable plants. Weed control or weed management, is a component or tactic of vegetation management, and the term Integrated Weed Management implies utilizing all methods of weed control in such a manner as to achieve optimum control with the least negative impact on non-target organisms and the environment. In this document we will assume this approach and use the term Vegetation Management Plan (VMP) to refer to site-specific management plans. Vegetation is a collective term that may mean weeds, native plants, invasive plants or landscape ornamentals, depending upon the situation and the context.

VMPs and Communities

Since VMPs are location specific, a number of factors should be considered when writing a plan. Geomorphology, topography, soils, site history, eventual site goals, and local community politics are some of the considerations. This analysis has an added dimension with public agencies or public land managers. Many public agencies are caught between support for a rapidly increasing population and an ever-shrinking budget. The effects of Proposition 13, passed by voters in 1978, have been felt by state and local agencies that are expected to tighten their spending but accomplish ever more work with a static budget. For example, communities that may be more aware of some of the controversy surrounding the use of chemical control methods may opt for a more custom approach to vegetation management. The question becomes how do agencies try to fairly meet the demands of their job responsibilities and public demands? Equity is a legitimate concern when agency budgets are being cut state and countywide. Add to this, a population that is accustomed to the infrastructure support paid by tax dollars, the demand for public agency accountability, and the crisis heightens.

Proposition 13

(excerpted from http://www.cbp.org/1997/9704pr13.htm)
"In June of 1978, California voters enacted Proposition 13 (and) reduced local property tax revenues by approximately $6.1 billion (53 percent) virtually overnight. Proposition 13 also made raising taxes more difficult by requiring state tax increases to receive the approval of two-thirds of the legislature and by imposing restrictions on the taxing authority of local governments . . . Proposition 13 fundamentally changed how public services are financed and administered at all levels of government in California. The relative prosperity of the 1980s enabled the State to assume a larger share of the cost of public services, particularly education. During the course of the State's repeated fiscal crises in the first half of the 1990s, a series of budget shortfalls led State lawmakers to shift costs back to the local level in order to balance the State budget. These actions pushed many local jurisdictions, particularly counties, toward fiscal crises . . . The limitations on local governments' ability to increase revenues

raise a number of issues in an era where "devolution" increasingly shifts programmatic and financial responsibility from the federal government to the states and from the state to local governments. Recent changes in welfare and other safety net programs for low income families, children, and the elderly, as well as broader efforts to balance the federal budget, all assume that states and localities are poised to take on new responsibilities. Yet in light of Proposition 13's impact on California, one must first ask whether California's communities have the fiscal resources to fulfill these new expectations."

Fairness

Public agencies need to balance needs across the entire geographic area under their jurisdiction as issues of social justice and environmental justice are important considerations in public and private policy and planning. As historically silent communities find their voices, they are requesting their fair share of resources and services from public agencies. "Issues of social justice, in the broadest sense, arise when decisions affect the distribution of benefits and burdens between different individuals or groups" (Clayton and Williams 2004). According to the U.S. Environmental Protection Agency (EPA), environmental justice refers to "the fair treatment and meaningful involvement of all people regardless of race, color, national origin, or income with respect to the development, implementation, and enforcement of environmental laws, regulations, and policies". Public agencies must keep the distribution of services in tension with a static to shrinking budget.

However, public agencies must fulfill their mandate to provide services and that typically involves vegetation management. For example, public infrastructure agencies must maintain the safety of roadways by pruning overhanging trees from roads and transmission lines, maintain flood control structures, and maintain defensible space for firefighters around structures. Public land managers must maintain safe open spaces for recreational purposes as well as improve habitat for wildlife. These responsibilities and more require vegetation management but within a finite budget.

In communities that have requested the use of a more expensive vegetation control method than one chosen by the managing public agency, a cost-sharing arrangement may be appropriate and equitable increased costs in one location means less funds for other locations. Tax assessment districts, volunteer programs, cooperative agreements and grants are cost-sharing options that allow communities choices for vegetation management and still meet the requirements of public agencies.

Climate Change, Drought and Weed Control

The effects of climate change are already manifesting in California with exceptional multi-year droughts and higher temperatures (Berg et al. 2015; Griffin and Anchukaitis 2014). While there is still uncertainty about the strength and timing of future impacts, it is likely that the need to secure local water sources will increase. One source of local water includes stormwater, rainwater that can be diverted, stored, or infiltrated. Use of planned stormwater facilities that detain and infiltrate rainwater are generally called "green infrastructure" or low impact development (LID) facilities. These facilities view stormwater as a resource rather than a nuisance. Facilities range from highly engineered filtration or separation devices to facilities engineered-as-natural ecosystems. These systems, such as, porous pavement, rain gardens and bio swales, rely on vegetation

FIGURE 1. Elmer Avenue Neighborhood Retrofit, a model green street project in Sun Valley, CA incorporates stormwater best management practices that capture and infiltrate rainwater from a 40 acre area.

and soils to filter and evapotranspire stormwater runoff *(Figure 1)*. The EPA has resources on their web site: *http://water.epa.gov/polwaste/green/*

Many publications summarize the benefits and advantages of green infrastructure and LID but too few mention weed management. Engineered facilities have maintenance instructions but engineered-as-natural ecosystems lack maintenance instructions. It is important that planners and engineers consider weed management methods in the planning stage and incorporate methods to ease maintenance and ensure long-term weed control.

http://water.epa.gov/polwaste/green/upload/bbfs2terms.pdf

https://www.casqa.org/resources/lid/socal-lid-manual

https://dpw.lacounty.gov/ldd/lib/fp/Hydrology/Low%20Impact%20Development%20Standards%20Manual.pdf

http://dpw.lacounty.gov/des/design_manuals/StormwaterBMPDesignandMaintenance.pdf

Why do we manage invasive plants?

Invasive plants, by definition, cause economic or environmental harm, or harm to human health (Federal Executive Order 13112, *http://www.invasivespeciesinfo.gov/laws/execorder.shtml#.UIHer2_AcZ4*). It is estimated that invasive species cost California at least $82 million in economic impacts and control costs *(www.cal-ipc.org)*. Researchers at Cornell University estimated damage and losses nationwide at $120 billion (Pimentel et al. 2005). These plants cause impacts to native habitat, water resources, soils, public health and safety, infrastructure, and can increase fire risk.

Invasive plants cause harm to native habitats by:

- outcompeting native plant species for water, nutrients, and space.

- reducing the amount of forage and physical habitat available to native wildlife.

- inhibiting native plant community recovery after damage due to fires, floods, storms, and human disturbance.

- altering aquatic and riparian systems by decreasing shade, and either causing erosion or conversely trapping sediments and clogging stream channels.

- in upland areas, forming monotypic stands that can fundamentally alter the three dimensional structure of communities. This may especially affect nesting birds that may need specific nest materials and specific nest platform structures.

Invasive plants impact water resources in several ways. By filling in stream channels, aquatic and riparian invasive plants can reduce channel capacity,

FIGURE 2. *Arundo donax* resprouting after a fire on the San Luis Rey River in San Diego County. Native trees are either dead, or still dormant. (Photo courtesy of Jason Giessow, Dendra Inc.)

increasing the likelihood of stream over-topping its banks during high flow periods. As mentioned above, riparian plants that form monotypic stands along banks may cause those banks to shear and fail at the root zone, sending biomass and sediment downstream further clogging the channel and potentially damaging infrastructure, such as bridges. Some invasive plants have higher rates of water use and evapotranspiration than native species, and may therefore reduce the water availability for either native habitat or for other uses.

Invasive plants may alter fire regimes. By their very nature, most species that are prone to invasion are weedy, and may grow vigorously when water is present, then die back when conditions become drier, leaving dry biomass behind – also known as "fuel". In desert communities (which are not adapted to fire), invasive annual grasses can grow in high densities, while native perennial species are more widely spaced with areas of mineral soil in between. When these annual grasses die back, they leave a drier, more continuous fuel load. This increases the risk of frequent fires, and the size and severity of fires when they do occur. In riparian systems, invasive species such as giant reed *(Arundo donax)* may also grow in monotypic stands at much higher densities than native species. When they die back, they again form a continuous fuel bed that can turn stream beds from a natural fuel break to a fire conduit.

Invasives can also interfere with recovery of landscapes after a fire. Native communities are disturbed by fire. Under natural circumstances, most communities are adapted to recover, and in fact this disturbance may be a beneficial way to "reset" environmental conditions if the fire frequency is not too high. However, when invasive species are present these plants may colonize or recolonize early and dominate the system *(Figure 2)*. As noted above, if "recovered" areas are now dominated by invasive species, they may be more prone to frequent fire, creating a cycle that allows for greater and greater spread of invasive species. The phenomenon of the replacement of native plant communities by invasive ones is called type conversion.

Chapter 2. Management Methods

Euphorbia terracina (terracina spurge) invading coastal sage scrub habitat near Malibu, CA.

There are four general approaches to vegetation management or effective weed management: prevention, eradication, control, and restoration (where appropriate). Techniques for limiting weed growth must be a part of every weed management program.

Prevention: The goal of prevention is to keep unwanted vegetation out of areas where it does not currently exist.

Eradication: This is the total elimination of a weedy plant from a site or area. Eradication methods commonly include a combination of approaches, including physical, cultural, biological, and chemical removal.

Control: Control includes actions taken to reduce or suppress weeds in specific sites or locations where eradication is not feasible. The goal is to eliminate or significantly reduce the damage done by the weed. Control methods include physical, cultural, biological, organic and chemical removal.

Restoration: The goal of ecological restoration is to return a site to a prior condition, typically one less disturbed by human influences. Restoration to this prior condition may be effective in preventing establishment of unwanted vegetation or in promoting beneficial vegetation.

PREVENTION

The three methods in this approach involve preventing the establishment of known or potentially invasive plants in an area where they do not exist. The methods include sanitation, quarantines, and education and outreach.

A. Sanitation

1. Using sources of planting materials (seed or nursery stock) that do not contain weed seed or live weeds.

2. Cleaning equipment before it is removed from a weedy area and moved into an uncontaminated area *(Figure 3)*. Designate sites for cleaning tools, vehicles, equipment, animals, clothing and gear before starting work. Preferred locations for cleaning are areas that are:

   - Already infested with invasive plants (with the assumption that you will leave clean)
   - Easily accessible for monitoring and control
   - Located away from waterways to avoid spreading invasive plants downstream
   - Consider the use of a portable cleaning station, especially when maintaining multiple sites with a variety of environmental conditions.

3. With livestock (goats, sheep, cattle): only using certified weed-free hay and forage; when animals leave a weedy area, placing them in pens for 2-4 days and feeding them clean hay before moving into clean areas so they evacuate their intestinal system of weed seed; and, inspecting their coats for weed seed. There are many other considerations with livestock grazing, such as disease spread to native sheep (bighorn and thinhorn sheep) as well as other native ungulates (elk, deer), and nitrogen loading and coliform excesses through feces (NPS 2016). *http://www.cal-ipc.org/ip/prevention/weedfreeforage.php*

4. Control weeds in adjacent areas so weed seed or other reproductive structures do not move into new areas. An example would be to control upstream infestations of giant reed *(Arundo donax)* first to prevent rhizome pieces from moving downstream during floods. *http://www.cal-ipc.org/ip/prevention/PreventionBMPs_LandManager.pdf*

5. Transportation corridors, recreation areas, trails, and other disturbed areas are often where new weed species first appear and where existing weed species spread the easiest.  These corridors include, but are not limited to highways, roads, riding and hiking trails, flood channels, bioswales, rivers, and streams. Control of invasive species in these areas can prevent spread into adjacent areas. *http://www.cal-ipc.org/ip/prevention/PreventionBMPs_TransportationUtilityCorridors.pdf*

FIGURE 3. Equipment is inspected and cleaned before leaving the work site to avoid seed transmission. (Photo courtesy of National Park Service).

Applicable to: All sites, especially those corridors mentioned above.

Pros: This method is very cost effective. Sanitation techniques seek to avoid transmitting problem plants rather than having to remove them. For example, since invasive species control in the U.S. is a concern of federal land managers, several federal agencies are in the process of adopting Weed-Free Feed policies to prevent the spread of exotic weed species through the activities of domestic animals *(http://www.fs.usda.gov/detail/okawen/recreation/horseriding-camping/?cid=fsbdev3_053645)*. The U.S. Forest Service and National Park Service developed Weed Free Feed rules to govern applicable park sites and applicable national forests.

http://www.fs.usda.gov/Internet/FSE_DOCUMENTS/fsbdev2_026443.pdf

http://www.nature.nps.gov/biology/invasivespecies/Prevention.cfm

Cons: Sanitation techniques are often logistically difficult. When a task is finished, it is easy to forget to do the necessary sanitation or to not make it a priority, particularly when resources are limited. Also, these methods can never be 100% effective; some weeds are always missed.

B. Quarantine

1. A state of enforced isolation or detention of a regulated plant to contain its spread. Quarantine applies to species listed by a state or federal regulatory agency *(Figure 4)*.

2. Federal law, enforced by U.S. Department of Agriculture Agricultural Plant Health Inspection Service, prohibits importation of plants listed as noxious at points of entry into the country. *https://www.aphis.usda.gov/publications/plant_health/2012/fs_imp_food_ppq.pdf*

3. California also has a noxious weed list aimed at preventing or containing listed weeds in the state that is enforced by California Department of Food and Agriculture (CDFA) and county agricultural commissioners. *https://www.cdfa.ca.gov/is/*

Applicable to: All sites.

Pros: Quarantine is very cost effective over the long term. It avoids problem plants by isolating or preventing them from becoming established.

Cons: Requires legal and regulatory authority that is subject to interpretation; is not typically adequately funded to be totally effective; relies on a 'dirty list', plants not listed cannot be quarantined. Not all plants are inspected due to, among other things, too many points of entry into the state. Plants can be difficult to identify and may be missed during inspection. Also staff limitations mean only a small sample of all material coming into the state can be inspected.

FIGURE 4. CDFA Border Protection Stations at California's border is our first line of defense in excluding invasive pest species. (Photo courtesy of LA County Agricultural Commissioner's Office).

C. Education and Outreach

1. Education and outreach are activities designed to promote public awareness and understanding of vegetation management issues and possibly change public behaviors that may influence vegetation management. For example, encourage people to plant native species or drought tolerant species in their yards and discourage people from activities that spread non-native species (surreptitiously planting "pretty" vines in public open space areas).

2. Examples of education and outreach activities related to vegetation management include:

   - Los Angeles County WMA's children's book, Invasive Weeds, What are they and why should we care about them? It explains the importance of controlling invasive weeds; is colorful, fun, and informative for both children and adults. *http://www.lacountywma.org/publications/childrens_book_on_invasives.pdf*

   - The California Invasive Plant Council's (Cal-IPC) "Don't Plant a Pest" brochure is an excellent example of an informative, education brochure. It explains the negative impacts of exotic species on wildland areas and discourages people from planting specific exotic invasive plants in their yards due to the potential for their escape. Escaped ornamental plants are a vegetation management issue for many landowners; if fewer people landscaped with these plants, this problem could be significantly reduced. *http://www.cal-ipc.org/landscaping/dpp/pdf/SoCalPrintable.pdf*

   - Cal-IPC and the Watershed Project have collaborated on The Weed Workers' Handbook, A Guide to Techniques for Removing Bay Area Invasive Plants (Halloran 2004). This handbook is geographically focused on San Francisco Bay area and discusses the worst pest plant species and guidelines for volunteer control projects. *http://www.cal-ipc.org/ip/management/wwh/pdf/18601.pdf*

   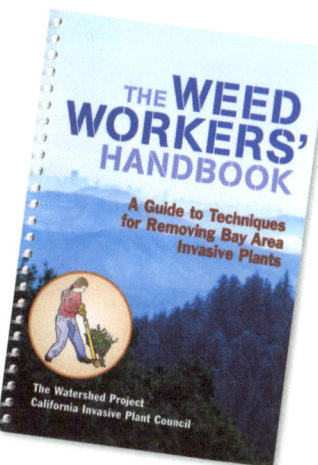

   - Plant Right is an organization that works with the nursery and retail gardening sector and experts to promote the use of non-invasive plant species in landscaping, and to provide education to the public. *http://www.plantright.org*

- The Los Angeles County Fire Department's program on fire safe landscaping provides information on vegetation management through education and outreach activities. Prevention of structure damage and insuring fire fighter safety are important issues in Southern California. Both fire fighter safety and structure protection are closely related to the management of vegetation surrounding homes. Due to this connection, the Los Angeles County Fire Department has a vigorous program of public education including site inspections, publications and web-based materials to educate the public on effective ways to reduce fire spread and increase fire safety in and around homes. *http://www.fire.lacounty.gov/forestry-division/forestry-fuel-modification/*

- The Care and Maintenance of Southern California Native Plant Gardens (O'Brien et al 2006) provides some basic control techniques for common garden weeds.

- Weed Control in Natural Areas in the Western United States (DiTomaso et al 2013) is a comprehensive resource for controlling over 240 species found in natural areas in the Western United States.

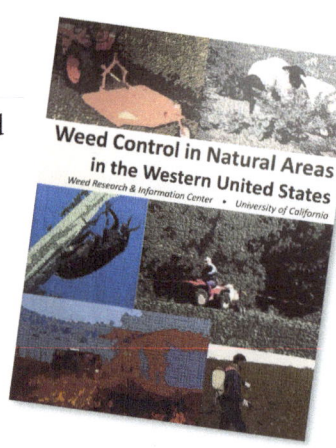

- For a wealth of downloadable weed education resources, see Cal-IPC's website. *http://www.cal-ipc.org*

Applicable to: All sites, although primarily those that have some public involvement.

Pros: Use of education and outreach strategies can raise public awareness and support for vegetation management issues. Further, raising public awareness and support can result in significant public involvement through volunteer work or other avenues. In addition, stakeholder interest and involvement may be vital to many vegetation management strategies, especially in the public arena. Education and outreach can increase such stakeholder interest and involvement.

Cons: As with any vegetation management strategy, development of education and outreach materials and techniques may be costly. You may need to weigh the costs of these strategies versus other more direct vegetation management actions. However, the benefits of education and outreach should not be undervalued as many unexpected benefits may come from such activities, such as, opportunities to build political and financial support.

ERADICATION

This is the total elimination of a plant from a site or area. Eradication methods commonly include a combination of approaches, including physical removal, cultural, biological, and chemical.

Applicable to: All sites except large holdings of land unless the weeds are isolated to small areas.

Pros: If species or infestations are identified early enough, it can be cost-effective as it eliminates a small problem before it gets large (this is often referred to as Early Detection/Rapid Response) *(Figure 5)*. It is often most effective when special emphasis is placed on transportation corridors. Examples in Los Angeles County are the eradication efforts directed against infestations of spotted knapweed *(Centaurea maculosa)*, halogeton *(Halogeton glomeratus)*, and alligatorweed *(Alternanthera philoxeroides)*.

Cons: This approach is used on a very limited basis because it is hard to accomplish, has to be maintained until there is no possibility that the weed will come back and therefore, can be very expensive. The timeframe for eradication may be three to five years. This method also requires a detection system staffed by well-trained people. Eradications almost always require quarantine and years of follow-up surveys. Example: the alligatorweed eradication program in Los Angeles County has been on-going for 30 years!

FIGURE 5. *Euphorbia dendroides* (tree spurge) eradication efforts at Chantry Flats, Angeles National Forest. (Photo courtesy of Los Angeles County Agricultural Commissioner's Office.)

CONTROL

Control includes actions taken to reduce or suppress weeds in specific sites or locations. The goal is to eliminate or significantly reduce the damage done by the weed, not necessarily to eradicate the weed.

There are four suggested methods to control vegetation: physical methods, cultural control, biological control, and chemical control. Each method may involve multiple techniques. All methods should be considered as part of a Vegetation Management Plan (VMP). This document addresses the following methods and their associated techniques:

Physical Methods
1. Hand pulling and hoeing
2. Fire
3. Flaming
4. Steaming/application of hot water
5. Foaming
6. Tillage (cultivation)
7. Mowing and shredding
8. Mulches
9. Soil Solarization
10. Structural
11. Weed Mats

Cultural Methods
1. Planting native or desirable plants
2. Living mulches and nurse crops

Biological Methods
1. Classical biological control
2. Grazing and other forms of herbivory

Chemical Methods
1. Selective vs. non-selective
2. Pre-emergence vs. post-emergence
3. Contact vs. systemic

Organic Methods

Other physical control methods were considered but were deemed inappropriate for inclusion in this document. For example, flooding is a physical control method that completely covers the land with water and is left for several weeks in order to kill the terrestrial plant species from anoxia. This method was not considered appropriate for L.A County as it is a waste of a dwindling resource and possibly creates habitat for invasive aquatic weed species. The LA WMA will consider other methods as they become available.

For land managers, one interesting weed control approach is the Bradley Method. In this approach, weed control is begun in portions of the site with the best stands of desirable native vegetation (those with few weeds) and proceeds slowly to areas with progressively worse weed infestations. The theory is that the smaller infested areas spread more rapidly than the older established infestation areas. Beginning control with areas of desirable native vegetation also protects your remaining resources rather than beginning in an area that has little resource value. For more information on the Bradley Method, see: *http://www.cal-ipc.org/ip/management/ipcw/mois.php*

For private residences, an excellent web resource for pest control is the UC IPM Online Statewide Integrated Pest Management Program, How to Manage Pests in Landscapes, Gardens, and Turf. *http://www.ipm.ucdavis.edu/PMG/menu.homegarden.html*

A. Physical Methods

Methods that uproot, bury, cut, smother, or burn vegetation. May have less potential to impact ground or surface waters than other methods. Since there are no residual chemicals, could be used in areas unsuitable to herbicides.

A.1. Hand Pulling and Hoeing

Applicable to: All sites except large holdings of land and transportation corridors.

Pros: Effective on annual species and some perennials, depending on growth stage. Does not require skilled labor but does require some initial education on weeds vs. desirable plants. People can get into areas that are not accessible with large equipment *(Figures 6 and 7)*. Necessary to achieve the level of control needed for highly visible landscape sites.

Cons: Not effective on many perennial species (perennials are injured by hoeing, but recover; many can be killed by repeated hoeing). Can be labor intensive and risky for workers along transportation corridors. Hoeing often creates significant soil disturbance, which lead to new opportunities for new weed infestations and the risk of soil erosion. Does not prevent weed regrowth of species that are able to re-sprout from root or rhizome fragments remaining in the soil. Poorly trained workers or volunteers can damage desirable vegetation by mistake.

FIGURE 6. Cal-IPC Student Chapter hand pulling milk thistle on Catalina Island. (Photo courtesy of Cal-IPC Student Chapter.)

FIGURE 7. *Arundo* removal in the Topanga Watershed. Volunteers use the "Hula Hoe" to recut *Arundo* stalks. (Photo courtesy of Jo Kitz, Mountains Restoration Trust.)

A.2. Fire

Fire can be used in a controlled, knowledgeable manner to burn off vegetation to achieve well-defined management objectives. Fire generally removes the above ground tissue, but may not affect perennial root systems so plants usually recover (e.g., which species are killed by a fire depends on many factors including the particular species involved, the fire intensity, and the season of the burn). If timed properly, burning can reduce seed production of many weeds, especially annuals. By knowing the relative fire tolerance of different plants, fire can select for desirable vegetation while reducing the undesirable plant species. Because of the risk of starting a wildfire, prescribed burns must be conducted under the direction of a certified fire-fighting authority *(Figure 8)*. Prescribed burns can provide training opportunities for state and local fire-fighters. *http://www.cal-ipc.org/ip/management/UseofFire.pdf*

FIGURE 8. Prescribed fire ignition. (Photo courtesy of Los Angeles County Fire Department.)

Applicable to: Limited to sites that can be burned safely and legally. Usually requires community coordination and specific weather conditions.

Pros: Can reduce or eliminate seed production by annual non-native forbs and grasses without significant damage to native perennials, but timing is usually critical to success. Removes accumulated duff that prevents the germination of native plants.

Cons: Can have serious safety risks to people, structures, wildlife, and natural areas if a burn is not properly assessed or conducted. Does not kill or significantly damage underground reproductive structures such as rhizomes, tubers, or seed. Requires a burn permit and strictly controlled environmental conditions. Can permit rapid re-invasion from seed bank, especially for fire-adapted invasives.

A.3. Flaming

FIGURE 9. Flaming treatment along paving stones.

Intense heat (2000°F) from a propane or butane torch is used to selectively kill succulent weed seedlings without harming woody native shrubs. Plants are not burned but "boiled" as water contained in plant cells heats and bursts the cell walls. The plant withers within a few minutes of flaming. Both hand-held and tractor drawn flamers are utilized.

Applicable to: Flaming is appropriate for sites under close control and typically small in size, such as private property, schools, urban areas. For example, it is appropriate for gravel, paving stones pathways and parking areas *(Figure 9)*. It is becoming a popular alternative to herbicide use on cropland. Is not appropriate in areas where organic mulch is being used as it often ignites.

Pros: Effective weed control of small succulent annuals and the seedlings of woody plants (such as scotch broom seedlings). Flaming can be cost effective. When used properly, flamers are quick, safe, and easy to use. Avoids soil disturbance.

Cons: Flaming is not effective on perennial plants or grasses (particularly crabgrass). Repeated flaming may be necessary for weedy grass control. There is a risk of ignition danger to standing vegetation (check with local fire department prior to investment). The cost can be greater than an herbicide, but may be lower than hand labor. Places workers at risk for burn injuries. Pressurized tanks of propane can be hazardous if handled carelessly.

A.4. Steaming/application of hot water

Steaming can be used selectively to kill succulent weed seedlings without harming other vegetation. It works best with new vegetation under six inches tall. Conversely, it does not work well with larger plants or those that have matured. Both handheld and tractor-drawn equipment are utilized *(Figures 10 and 11)*.

Applicable to: Sites under close control and small in size, such as private property, schools, urban and agricultural areas.

Pros: Steaming is effective weed control of small seedlings and annual plants; cost effective for small plants and when areas to be treated are small in size or in strips accessible with tractor-drawn equipment. Can be a very selective method, only killing weeds and not damaging other vegetation unless the weeds and desirable plants are closely mixed. No soil disturbance. No risk of inadvertently starting fires. Requires no permits or licenses.

Cons: Does not control perennial plants, large grasses, or plants with extensive underground roots or rhizomes. The amount of time required for this work is equivalent to hand labor but the cost is slightly higher due to the initial cost and maintenance of the equipment. Places workers at risk for burn injuries from equipment and back splashing from extremely hot mud. Full safety gear is necessary including hearing protection.

FIGURE 10. Weed control using steam by hand. (Photo courtesy of Dr. Thaddeus Gourd, Colorado State University Cooperative Extension for Adams County.)

FIGURE 11. Weed control using steam by tractor. (Photo courtesy of Dr. Thaddeus Gourd, Colorado State University Cooperative Extension for Adams County.)

A.5. Foaming

Pressurized application of hot surfactant foam that comprises a biodegradable mixture of corn and coconut sugar extracts and superheated steam for steam-killing vegetation. The mixture is considered an "organic," naturally occurring compound and is, therefore, not regulated as an herbicide by U.S. EPA. Foaming can be used selectively to kill succulent weed seedlings without harming other vegetation. Both hand-held and tractor-drawn equipment are available.

Applicable to: Sites under close control and small in size, such as private property, schools, and urban areas where complete vegetation removal is sought. Roadside application is feasible with the truck mounted mechanical boom delivery system. Foam cannot be used near surface water and a concentration of 3 mg/liter can be toxic to fish. The effects of foam need to be fully studied.

Pros: Foaming is effective weed control of small seedlings, annual plants and some perennials (Figure 12). Can be a very selective method, only killing weeds and not damaging other vegetation unless the weeds and desirable plants are closely mixed. No soil disturbance. No risk of inadvertently starting fires. Hot foam requires no permits or licenses since the California Department of Pesticide Regulation (DPR) determined that hot foam is not a pesticide. Can be used in varying weather conditions including light wind and rain without fear of pesticide drift or residue.

Cons: Currently systems are only available by lease. Some perennials, especially plants with extensive underground roots or rhizomes, may require more than one application to attain full control. The amount of time required for this work is equivalent to hand labor but the cost is slightly higher due to the equipment lease and cost of foam. Uses water quickly so a nearby water source is recommended. Safety gear is recommended including eye and hearing protection. The foam can cause eye irritation.

FIGURE 12. Ellen Mackey applies super-heated foam to bermuda grass at the Mountains Recreation Trust property during a pilot test.

A.6. Tillage (cultivation)

Tillage includes practices such as roto-tilling, disking and plowing that disturb the soil. In the process, plants are cut off at the soil line, uprooted, or smothered *(Figures 13 and 14)*. There are two basic types of tillage, vertical and horizontal. Vertical tillage is done with a disc or harrow that cuts down through the soil. This type of tillage is cheaper because the implements are easier to pull through the soil and they disturb the soil less that horizontal, but they do not sever root connections of perennial plants very well. Horizontal tillage utilizes plows, sweeps, and roto-tills. It does sever roots well, but is very expensive and makes soil more prone to erosion.

Applicable to: Sites accessible to motorized equipment (tractors, roto-tillers).

Pros: Tillage is very effective, quick, and can be less expensive than other methods that have similar results. Will kill perennial plants if done routinely.

Cons: Tillage requires skilled or semiskilled labor and supervision, and equipment that is expensive to purchase and maintain. A significant source of erosion in many sites. Dust can create a vision hazard along highways. Horizontal tillage can bring buried weed seed to the soil surface where it can germinate. Rhizomes of perennial weeds can be cut up, which can lead to more individual plants and these pieces can be spread on equipment. Heavy discs can damage underground water and gas pipes, etc. (check *www.digalert.org*). The repeated disturbance from tillage may be more environmentally damaging long-term.

FIGURE 13. Discing being used to control weeds in a firebreak in the Antelope Valley, Los Angeles County in 2002.

FIGURE 14. Disked land in White Point Nature Preserve, grassland restoration project. (Photo courtesy of Palos Verdes Peninsula Land Conservancy.)

A.7. Mowing and shredding

Mowing, cutting, and shredding weeds with motorized equipment or hand operated tools such as string trimmers, chain saws, brush cutters and slope mowers. Heavy equipment and hand-held tools are discussed together then separately. When chipping wood in areas with suspected Polyphagous Shot Hole Borer (PSHB) infestation, woody debris should be chipped smaller than 1". *http://ucanr.edu/sites/socaloakpests/Polyphagous_Shot_Hole_Borer/*

or hand held tools. The results of a study evaluating different control methods associated with fire lines, including several different types of mechanical control, can be found at *http://www.werc.usgs.gov/OLDsitedata/fire/seki/ffm/.*

FIGURE 16. Flail mowing *Brassica nigra* (black mustard) at White Point Nature Preserve in San Pedro (2002). (Photo courtesy of Palos Verdes Peninsula Land Conservancy.)

FIGURE 15. Hydroaxe machine cuts and mulches *Arundo donax* at Whittier Narrows. (Photo courtesy of Bill Neill, Riparian Repairs.)

Applicable to: Sites accessible to motorized equipment or by individuals using portable tools. Both heavy equipment (riding mowers, tractor pulled implements, etc.) and hand held tools (string trimmers, swing blades, chain saws and brush cutters) can be used to remove unwanted vegetation. The scale of project may determine which equipment is chosen, but consider the following information if there is a need to choose between heavy equipment

Pros: Mowing and shredding is aesthetically pleasing and more efficient than manual labor. It reduces fuel for fires, reduces seed production (if performed before the plant goes to seed), and lowers water use. Can be used on dense vegetation and does not disturb the soil surface.

- *Heavy Equipment* - Heavy equipment can be very efficient, covering large areas in short amounts of time *(Figure 15 and 16)*. In addition, use of such equipment by a professional operator is quite safe on the proper terrain (not too steep or rocky). The operator is typically shielded from certain environmental hazards such as snakes, throwing rocks and extreme heat by the vehicle being operating.

- *Hand Held Equipment* - Use of hand held tools such as brush cutters or weed whips allow a high level of control over vegetation management. Operators can work around inclusions of desirable vegetation, animal burrows, etc *(Figure 17)*. There is only minimal impact to the soil due to the small weight of the individual operator versus a piece of heavy equipment. Considered good for use in sensitive areas.

FIGURE 17. Rio Hondo Fire Academy crews use chainsaws to remove *Ailanthus altissima* (tree of heaven) within the Monrovia Hillside Wilderness Area. (Photo courtesy of Rachel Wing.)

Cons: Results are species specific. Safety to operators and bystanders is a concern. Equipment can throw rocks and debris into traffic lanes and onto private property, can start fires, and create visibility problems from dust. Leaving chipped woody debris can promote the spread of plant pests and diseases such as various bark beetles. *http://firewood.ca.gov/*

Equipment is expensive to purchase and maintain, requires skilled labor and supervision. Debris removal is expensive. Noise can bother local residents or wildlife. Can encourage weeds if improperly timed. May not be as cost-effective as chemical application. Weeds with seeds (especially invasive weeds) should be mowed before seed matures or removed by hand before mowing takes place.

- *Heavy Equipment* - Depending on the skill of the operator and the site being treated (are desirable plants intermixed with unwanted vegetation?), use of heavy equipment may result in damage to resources (such as desired vegetation, soil compaction or damage to small animal burrows).

- *Hand Held Equipment* - The use of hand held equipment may involve more labor costs and longer time to complete large projects. In addition, individuals using brush cutters or weed whips may be exposed to environmental hazards such as loose rock, snakes, poison oak and heat.

A.8. Mulches

Mulching is a simple, relatively inexpensive method of controlling weeds by spreading a protective layer of material on the ground that effectively reduces weed growth by excluding light from the soil *(Figure 18)*. Mulch materials can be organic (compost, manure, bark chips, newspapers, straw, hay, sea weed mulch and pine needles) or inorganic (rocks, gravel, carpet padding, plastic sheeting, landscape fabrics, ground rubber tires). In addition to light deprivation, organic mulches can tie up available nitrogen through decomposition and starve weed seedlings as they attempt to grow. Timing for application is important and depends on the objective. Mulches stabilize the soil temperature by providing an insulating barrier between the soil and the air. Summer soil temperatures will be cooler than in the adjacent un-mulched soil. Winter soil temperatures will not get as cold and will warm up more slowly in the spring and cool down more slowly in the fall than un-mulched soil. The desired soil temperature becomes an important consideration if mulching is a primary method for weed control in an area that is also a planted native landscape. Also mulching too early can delay soil drying and subsequent root growth that is dependent upon sufficient oxygen content in soil and reasonably warm temperature in the root zone.

Applicable to: Can be used in a variety of sites, but different sites will limit the mulch material that can be used. To be effective, mulch should be at least 3 to 4 inches thick.

Pros: Mulches are generally very effective on annual weeds, but less so on perennial weed species. Recycled materials such as tires, plastics, papers, wood chips, and compost can be used as mulch. Have additional benefits such as conserving soil moisture, maintaining even soil temperature, reducing soil compaction, adding nutrients. Mulches can also add a "finished" look to the landscape. Pine needles can increase the acidity of soil around acid-loving plants such as rhododendron or azaleas.

FIGURE 18. Los Angeles Conservation Corps crews spread mulch to help control weeds at a new rain garden installation.

Cons: Organic materials can ignite. Improper placement or site selection can result in clogged water runoff conveyances or drains. Labor intensive to install and maintain and does not work well on perennial species. Some materials are expensive. Mulches, such as hay and straw, work well but may harbor weed seeds. Unless mulch is weed-free, it can introduce new invasive weeds to an area. The moister, cooler environment can be very attractive to other pests, such as earwigs, slugs and sow bugs. Excess mulch, particularly if applied right against the stem or trunk of landscape plants, also leads to root crown death, conditions favorable for disease development, and plant death. Organic mulches can change the soil structure and enrich soil to the detriment of native species. When possible in restoration projects inorganic mulches should be given serious consideration.

A.9. Soil Solarization

Soil solarization is a simple non-chemical method that uses clear plastic to trap heat energy from the sun to bring about physical, chemical, and biological changes in the soil that will kill soil pathogens and weed seed *(Figure 19)*. The treatment area is watered if dry, completely covered with clear plastic tarp that is left in place 4 to 6 weeks. The top 6 inches of soil will become 20-40 degrees higher than ambient temperature when solarization is done properly. Solarization is most effective when done in July, August and September when days are warm and sunny. See *http://www.ipm.ucdavis.edu/PMG/PESTNOTES/pn74145.html* for more information on solarization.

Applicable to: Can be used in a variety of sites.

Pros: Solarization is a non-chemical method of soil disinfestations. Provides safe and effective control of weed seed and plant pathogens to a depth of 6 inches if done correctly with sufficient radiant heat energy from the sun.

Cons: Labor intensive to install and maintain and does not work well on perennial species. Some materials are expensive. Plastic disposal is an issue. An ultraviolet resistant plastic must be used or the plastic will degrade into thousands of small fragments before the process is complete; this plastic is not readily available. Solarization does not work well in coastal areas of L.A County because of more moderate temperatures and cloudy conditions. Solarization can kill beneficial microbes and insects as well as any native seeds in the seed bank. Effective only at warm times of the year so timing is critical and may cause delays in poorly scheduled projects.

FIGURE 19. Solarization technique used in Fay's Wildflower Garden at Rancho Santa Ana Botanic Garden. (Photo courtesy of Michael Wall, RSABG.)

A.10. Structural

FIGURE 20. Structural approaches can be highly decorative and, as a hardscape element, reduce the need to continuously add organic mulch, lower maintenance costs, and infiltrate rainwater.

Designing facilities and landscapes with the intention of minimizing the future need for vegetation control *(Figure 20)*. A maintenance plan including weed management should be included as part of the design.

Applicable to: Especially appropriate for new sites, renovations, linear parkways, or roadways.

Pros: Use of hardscapes, thoughtful installation of irrigation systems, banking hardscape surfaces to divert stormwater to plantings, and careful selection and spacing of appropriate plant materials can reduce maintenance requirements, watering, and expenses.

Cons: Initial cost is high and design flaws can be expensive to retrofit or repair.

A.11. Weed Mats

Weed mats are a specialized type of mulch or ground cover, usually porous plastic, that covers the soil to stop weed growth without chemical use but allows water to move through to soil *(Figure 21)*. Weed mats vary in thickness and durability. Some resistant to UV radiation with life spans of at least 15 years.

Applicable to: Under guardrails along highways, fences, signs, utility poles, hydrants or anywhere the normal weed control method would be weed-whipping or spraying herbicides. Can be used in low impact development projects that transition from developed hardscaped areas to green spaces.

Pros: The right product can control weeds, reduce runoff, long-lasting, can be maintenance free, pesticide free, can conform to any shape, and can be fire-resistant.

Cons: Initial cost can be higher but the right product can be very cost effective over time. If weed mat fabric is too thin, it can require replacement in several years and fabric will end up in landfill.

FIGURE 21. Weed mats underlying highway guard rails in Northern California. (Photos courtesy of Russ Mason, U-TECK.)

B. Cultural Methods

These are not weed control methods as such, rather they are practices that favor desirable vegetation so it can outcompete undesirable plant species.

B.1. Native Planting

Planting native or desirable plants in a way that provides them the greatest likelihood of establishment, thus increasing their ability to compete with the weeds *(Figure 22)*. To minimize weed encroachment after installation:

- minimize soil disturbance

- minimize fertilizer use which favors invasive plant growth

- adding arbuscular mycorrhizae or inoculating with native weed-free soil may foster growth of native plants

- irrigate less often and deeper to reduce opportunities for invasive plant germination; allow soil to dry down in the top 6 to 12 inches of soil prior to irrigating again (depending on plant growth and root establishment).

Applicable to: New or renovated landscaping, bio-swales, rain gardens, and roadside and parkway sites. Irrigation systems that are designed to provide for the seasonal water needs of natives are often a key to success.

Pros: Long-term vegetation control will be reduced if the desirable vegetation grows quickly and fills in the landscape.

FIGURE 22. National Park Service staff planting native vegetation in a disturbed area at Paramount Ranch. Plantings are part of a restoration project to replace non-native perennial pepperweed *(Lepidium latifolium)* with native plant species found in similar areas in the Santa Monica Mountains.

Cons: Requires landscape designs that will be successful, this is not always well understood. Some California native plants may not compete well with invasive plants; vigilant weed maintenance may be required to keep invasives from getting established and desirable plants from being overrun during plant establishment (2 to 5 years post-installation depending on environmental conditions).

B.2. Living mulches, cover crops, and nurse crops (Fennimore and Bell 2014, Donaldson et al 2002)

These terms refer to the practice of planting an aggressive annual plant in conjunction with or before planting a desirable species to prevent weed problems. These methods are used in agriculture, especially in orchard crops. The theory is that the annuals compete with the weeds, then die and leave the space to the desirable shrub or tree. Species that may be used for erosion control and nurse crop species include wooly plantain *(Plantago insularis)* and small fescue *(Festuca microstachys)*. Other native species that are good competitors include chaparral mallow *(Malacothamnus fasciculatus)*, giant wild rye *(Elymus condensatus)*, and California blackberry *(Rubus ursinus)*.

Applicable to: New or renovated sites, preferably with some irrigation. Also, extensively used in agriculture.

Pros: Low cost if successful.

Cons: There is not much experience with this method. There are limited examples outside of agriculture. As stated above, the living mulch can become a weed, especially if it persists longer than expected or reproduces.

C. Biological Methods

Using a living organism to manage the population of a plant pest species.

C.1. Classical biological control

Most invasive plants are non-native, meaning they come from distant countries, usually a different continent. When they arrive in California, they do not bring their natural enemies (the organisms that feed on/control them back in the country of origin) with them. A search can be made for a natural herbivore of the weed, typically an insect, which can be brought into the U.S. for release on the weed if certain standards can be met *(Figures 23 and 24)*. The principal standard is that the insect bio-control agent cannot feed on valuable crops, ornamental plants, or native plants. Biological control is not intended to eradicate a weed, just to suppress it to acceptable levels.

Applicable to: Most applicable to open areas where other methods of control are too costly and where less than 100% control is acceptable.

Pros: Very cost effective when it is successful and it is self-sustaining; the bio-control agent seeks out the host plant.

Cons: Not generally applicable to any plant species that has a lot of related species growing in the state, especially native species. Expensive up-front costs to find the control agent, conduct required testing, and releasing the agent. Time intensive process can take over a decade to determine specificity and safety. Initial and often very extensive research must first be done by government agencies such as the California Department of Food and Agriculture and/or USDA. Typical control of a successful project is between 60-90%; this is not acceptable for all weeds, sites or habitats.

FIGURE 23. Photos showing a healthy yellow starthistle flower (left) and one parasitized (middle) by the hairy weevil *(Eustenopus villosus)* (right). The weevil was introduced into Los Angeles County by the County Agricultural Commissioner Department in 1998 in an effort to control the spread of yellow starthistle. (YST photos courtesy of Ray Smith, LA County Ag Comm Office; Weevil photo courtesy of Eric Coombs, Oregon Department of Agriculture, *www.forestryimages.org*)

FIGURE 24. Dead tamarisk *(Tamarix* spp.) on the Colorado River near Moab, Utah that was defoliated by the tamarisk leaf beetle *(Diorhabda* spp.). (Photo courtesy of Bill Neill, Riparian Repairs.)

C.2. Grazing and other forms of herbivory

This is generally grazing with sheep, goats, or cattle, but can also include plant-eating fish such as tilapia and grass carp. This type of control is analogous to mowing because it generally only removes the top parts of plants but not roots or rhizomes *(Figure 25)*.

Applicable to: Limited to sites that are accessible to animals and where they can be fenced or managed.

Pros: Can be cheap and effective for control of annual plants and to suppress seed production. Grazing is analogous to mowing for perennial weed control.

Cons: Animals have to be managed for effective control and to prevent the animals from eating desired vegetation. The animals have to be protected from predators and dogs. Safety is a concern on highways. Nutrient addition from feces can make the site more hospitable to invasive species. Some herbivores may not be selective and may damage non-target species. Some plants are poisonous to grazers. Animals must be managed to assure that they do not transport exotic species from other areas either in feces or on their fur.

FIGURE 25. *Top:* Goats being used to reduce a potential fire hazard posed by weeds and brush in Claremont, CA. *Middle:* Before grazing. *Right:* After grazing. (Photos courtesy of J. Lopez, Los Angeles County Fire Department.)

D. Chemical Methods
Modern herbicides are typically organic molecules derived from petroleum.

FIGURE 26. *Centaurea maculosa* (spotted knapweed) treatments in the San Dimas Experimental Forest. (Photo courtesy of Los Angeles County Agricultural Commissioner's Office.)

Applications in public situations are performed by trained applicators *(Figures 26 and 27)*. Some of the applications may also require a written recommendation from a Pest Control Advisor (PCA). *Appendix A* discusses the regulations associated with the use of chemical control measures.

Some herbicides are available that are approved for use in organic agriculture; these are generally weak acids like vinegar. Herbicides fall into a wide and diverse group of classes, so they are generally categorized by use patterns, as listed below. Any given herbicide can be any combination of the three categories listed below.

a. <u>*Selective vs. non-selective*</u> – Selective herbicides are the larger group and refers to chemicals that kill some species of plants but not others. Non-selective herbicides, like glyphosate, kill any plant, but this is not absolute. Some plant species are harder to kill than others (e.g., require higher rates or dosages for effective control), annuals vs. perennials for example and herbaceous species vs. woody species.

b. <u>*Pre-emergence vs. post-emergence*</u> – Pre-emergent herbicides are applied to the soil and kill weed seedlings as they germinate. Post-emergent herbicides work on emerged plants by interfering with physiological processes in the plant.

FIGURE 27. *Spartium junceum* (Spanish broom) treatments in the San Dimas Experimental Forest. (Photo courtesy of Los Angeles County Agricultural Commissioner's Office.)

c. <u>Contact vs. systemic</u> – Contact herbicides only damage the tissue they are applied to, killing plants by desiccating leaf and stem tissue. Systemic herbicides enter into the plant and move to leaves, stems, or roots and have their effect on physiological processes at that location.

Applicable to: A wide variety of sites of private or public entities that can control access during and after herbicide application. Environmental considerations are very important, such as wind speed, proximity to schools or housing, potential for offsite movement in dust or water, and proximity to habitats for protected animal and plant species.

The California Department of Pesticide Regulation (DPR) has established Groundwater Protection Areas (GWPAs) to prevent further groundwater contamination in areas where pesticide contamination has occurred. When considering chemical control methods, individuals and agencies in the counties of Los Angeles, Orange, and Riverside should check for the appropriate regulations and locations. *http://www.cdpr.ca.gov/docs/emon/grndwtr/*

Pros: If used properly, herbicides can minimize exposure of personnel to vehicle traffic, exposure of the public to equipment and traffic diversions that might be required for mowing or burning. Herbicides can selectively control undesirable vegetation and leave desirable vegetation unharmed. Relatively inexpensive and effective compared to many other methods.

Cons: Inappropriate application of herbicides (e.g., during excessive winds that move the herbicide off the target area, during periods of the day when people are normally present, having faulty equipment that results in leaks or spills) can harm animals and cause damage to the environment. Conflicting information regarding environmental toxicity is readily available online. In certain application areas, public perceptions of herbicides will need to be addressed. Cal-IPC's herbicide BMP manual presents ways land managers can protect wildlife when using herbicides to control invasive plants. *http://www.cal-ipc.org/ip/management/BMPs/index.php*

E. Organic Methods

Synthetic and non-synthetic pesticides that are allowed for use under the USDA National Organic Program standards; some are derived from natural sources not synthetically manufactured.

FIGURE 28. OMRI logo Products that meet organic standards under the USDA National Organic Program are allowed to display the OMRI Listed® seal.

Debate concerning the safety, estrogenic effects, synergistic effects and efficacy of synthetic herbicides will continue until more research on these products is conducted and published. In the meantime, maintenance crews of some public landscapes are asked to maintain parks, roadsides, fields, and detention basins without the use of pesticides. However one feels about the safe use of pesticides, the public is becoming alarmed about the reported effects of pesticides derivatives on aquatic species and surface/groundwater. A number of alternatives have emerged in a thriving market of organic/non-synthetic chemicals for use in agricultural and public landscapes. However, finding substitutions for petroleum-based/synthetic herbicides is not easy and reliable information is sparse.

OMRI, the Organic Materials Review Institute, a 501(c)(3), reviews pesticides against the USDA National Organic Program (NOP), Organic Production and Handling Requirements. Those products allowed by OMRI are listed in the OMRI Products List© and eligible to display the OMRI Listed® seal *(Figure 28)*. Since OMRI program participation is voluntary, products not assessed or listed by OMRI may still meet USDA NOP rules. A product's absence from the OMRI Lists does not necessarily mean it does not comply with NOP standards.

The list of weed controls vary widely from plastic mulch to weak acids to mined minerals to plant extract derivatives to coconut fiber to biocontrols. Since "natural" or organic does not mean non-toxic, OMRI Listed products can be harmful to the environment, humans, and wildlife if improperly used. Therefore, all material/products should carefully follow guidelines, restrictions and regulatory annotations as well as label instructions on products. *Appendix C* discusses the regulations and organizations associated with organic control measures.

Applicable to: Projects installed with stormwater infiltration systems (Low Impact Development projects) or communities with chemical sensitivity issues that require the use of non-synthetic herbicides or organic methods of weed control.

Pros: These products are non-synthetic pesticides, derived from natural sources not synthetically manufactured. Though we recommend that all applications in public situations are performed by trained applicators following an IPM approach, organic pesticide applications do not require a written recommendation from a Pest Control Advisor (PCA).

Cons: Published studies from UC Davis and Riverside suggest that organic pesticides are not as effective in crop land situations where agriculturists wish to treat with a pesticide at infrequent intervals. To date, no studies on the effectiveness of pesticides in non-crop situations, most especially wild lands, have been conducted. This remains an area for further research. However, organic farmers use a number of methods to control pests including: mechanical (tillage), cultural (cover crops), biological controls (living mulches), and chemicals (rotenone and pyrethrin). Field situations may require a combination of methods with shortened treatment timing to accomplish the desired weed control. The products are not registered or reviewed by the EPA for impacts to human and environmental health (short or long-term impacts). This is problematic because many chemicals can do harm especially when used improperly. Decisions should be made by a certified Pesticide Control Applicator.

OMRI lists products that are allowed in organic agricultural applications whereas we are seeking solutions to open space/wildland, water-quality, and wildlife sensitivity issues.

OMRI does not investigate the effect of listed pesticides on non-target species including endangered, threatened, or sensitive species. In addition, OMRI does not assess synergistic or cumulative effects of listed pesticides.

RESTORATION

Restoration is defined as "the process of assisting the recovery of an ecosystem that has been degraded, damaged, or destroyed" (SERI, 2004).

The goal of ecological restoration is to return proper ecological functioning to a degraded ecosystem, typically one less disturbed by human influences. Restoration may be effective in preventing establishment of unwanted vegetation or in promoting beneficial vegetation *(Figure 29)*.

Applicable to: Numerous sites. The scale, goals, and timeline of restoration projects can vary widely depending on the site and site-specific management goals. Ecological damage present at a site may include disruptions of ecological processes or may be limited to small-scale disruption of ecological features or something in between these two extremes. Examples of altered ecological processes include decreased water availability due to high water demand and groundwater pumping, polluted water and increased nutrients in water, increased fire frequency, and increased soil erosion. Examples of small-scale disruptions include soil disturbance, a road or trail, presence of a barren area due to an old homestead or new development, barren areas cleared for fire breaks near structures, and introduction of non-native plant species due to past or present human activities. These changes in ecological features may or may not reflect underlying changes in ecological processes. For instance, a site may have an area dominated by non-native weedy grasses which is a change in ecological condition, however, this change may be due to an alteration in ecological processes such as increase fire frequency or land use change.

FIGURE 29. PVPLC staff planting natives at Palos Verdes Nature Preserve. (Photo courtesy of Palos Verdes Peninsula Land Conservancy.)

Once an ecological problem has been identified at a site, a plan to restore the area to a former, less degraded, ecological condition is developed. This plan may include actions such as grading an area to restore the natural contours, removal of surface water impoundments, use of prescribed fire to re-instate a more natural fire return interval, invasive species control actions, and planting of native species found in similar habitats.

Ecological restoration can play several roles in an integrated vegetation management program. For some sites, ecological restoration itself, returning a site to a former, less degraded condition may be the goal. In other cases, ecological restoration may be a tool used to achieve or maintain other vegetation management goals. For example, if your goal is to keep invasive species out of a particular area, one way to achieve this goal is to densely plant the area with native vegetation. In this case, the goal is not the restoration itself, but a weed-free condition. The weed-free condition could be achieved by other means (constant treatment of invasive species to keep them out of the site and prevent their reproduction) but revegetation with native species may be one way to achieve your stated site goal.

Pros:
- Habitat for native wildlife.
- There is an increased potential for volunteer site stewardship with a diverse list of jobs beyond the initial planting such as watering and maintaining weed-free areas.
- It can be cost-effective in the long term due to reduced irrigation requirements.
- Compared to ornamental/non-native plants, natives enhance roadside water quality and roadside visual quality while requiring less maintenance with little irrigation requirements once they are established due to slower growth patterns.
- Erosion control.
- Increased biodiversity.
- Some native plants may reduce the risk of wildfire.
- Opportunities to educate the public and surrounding community about the importance of habitat restoration and other local environmental issues.

Cons: This approach is used on a limited basis because each project requires site-specific goals; plant material can be limited and may need to be grown specifically for the project. Contract growing makes this method potentially more expensive, especially short-term. Restoration almost always requires years of follow-up monitoring by well-trained people. However, with some effort and training to recognize native plant species, collect mature seeds, learn to propagate, plant, and care for newly planted plants, native plant material can be easily obtained for the restoration site.

Methods of native community revegetation

Develop an enhancement plan. Recognize native plant species that belong in the restoration site's habitat. Start with identifying the native plants near the restoration site. Develop a list of native plants to use, consider the native species and plant communities growing nearby, and easy to propagate native species. Identify and properly collect mature seeds. Consult books in references for seed propagation techniques.

If using container plants, or plants grown from seed, start the propagation well in advance for planting into the site the next year.

Plan where each container will be planted, keeping in consideration how the new plants will be watered; planting near a source of water, or the riparian corridor, is best but not always possible. Plants will require irrigation through their first year (or more under drought) until establishment and then irrigation will no longer be needed. Consider installing a temporary, above-ground pipeline made with small, non-pressurized PVC pipe connected to a water source. Through the use of valves, water can be transported fairly long distances cheaply. To water, connect hoses to the valves and hand water. Other methods include a large water tank on the back of a truck, the use of buckets and other containers to hand water. Planting and establishing species first that will provide shade later to other plants may help reduce watering needs.

Plant during the rainy season, which generally occurs in Southern California from November through March. The following planting method was derived from the Tree of Life Nursery website *(www.californianativeplants.com)*:

- Dig hole twice as deep and twice as wide as plant container. Break up large clods and try to avoid the smooth-sided "bathtub" effect in the hole *(Figure 30)*.

- No soil amendment is recommended for the native plantings. The soil should be soft and friable. Eliminate large rock and clods from the backfill soil.

- Fill planting hole with water and allow percolation (draining) into subsoil.

- Spill some backfill material into the bottom of hole, moisten, tamp and mound slightly.

- If installing a cage for the new plantings, place cage in hole with approximately 10 inches below ground and 10 inches above ground.

- Add more backfill above bottom of cage. Set plant root ball atop the moistened backfill so that plant collar is one-inch higher than finished grade.

- Fill in backfill around rootball with friable soil. Be sure collar is still higher than grade. Any rocks dugout of the soil should be re-buried one-third in depth into the terrain, either just outside the rootball or outside the watering basin.

- Create a watering basin, in the shape of a donut, with the bottom ring of the basin outside the rootball, using remaining excavated soil.

FIGURE 30. AmeriCorps Team planting natives at Palos Verdes Nature Preserve. (Photo courtesy of Palos Verdes Peninsula Land Conservancy.)

FIGURE 31. *Amorpha fruticosa*, the host plant for the state butterfly, is easily grown from 6' pole cuttings in long pots. The natural fencing on the right comprises several hundred such rooted cuttings.

- Apply 3-6" of organic weed and disease-free mulch, topdressing outside the watering basin and lightly inside the basin, with no mulch covering the plant's root crown.

- Irrigate thoroughly, filling the basin with water and sprinkling around to settle backfill, mulch and berm. Allow to soak in and repeat.

Maintain a regular watering schedule for at least the first year.

Other revegetation techniques include:

- <u>Pole cutting</u> - Revegetating with live pole cutting stakes is another technique for revegetation of a restoration site. Advantages include not having to propagate plants in a nursery, can be immediately planted, and use little water *(Figure 31)*. Good species for pole cuttings include willows *(Salix* spp.*)*, mulefat *(Baccharis salicifolia* subsp. *salicifolia)*, false indigo-bush *(Amorpha fruticosa)*, coyote brush *(Baccharis pilularis)*, elderberry *(Sambucus nigra* subsp. *caerulea)*, cottonwoods *(Populus* spp.*)*, mugwort *(Artemisia douglasiana)*, California blackberry *(Rubus ursinus)* and California wild rose *(Rosa californica)*.

  Sprigs are best cut while in deepest dormancy, generally December-February. The best pole cuttings are generally young straight branches which commonly occur as epicormic sprouts in the internal structure of mature willow trees. All cutting material should be collected from healthy trees where the cutting of vegetation would the plant. Look for healthy, green branches of desired species and cut at the node; a desirable size is approximately 0.5-3 inches in diameter and 1.5-4 feet in length. Cut the bottom end at a 45-degree angle to assist in installation and for easy directional identification. Cuttings should be stored in water containing a diluted solution of a rooting auxin until the sprigs are installed in the soil. A pilot hole may be made using rebar or a digging bar, and the stakes driven into the soil up to 90% of their length. Stakes driven in with a mallet will require a fresh top cut.

- <u>Direct Seeding</u> - Advantages include little preparation other than collecting seeds and distributing. Only a few, specific species can be grown this way and often include the disturbance-loving native plants that grow readily along roadsides and near disturbed areas. Disadvantages include the fact that many species cannot be successfully grown from throwing seed onto bare soil; seeds have a high predation rate.

Chapter 3. Components of a Vegetation Management Plan

A **Vegetation Management Plan** (VMP) should include:

1. The need or reason for vegetation management – i.e., a specific need or target problem species
2. The areas and acreage to be treated
3. Method(s) of treatment and reasons for method(s) at the particular site
4. Timeline for treatment
5. Follow-up and monitoring

Examples of a Roadside and Wildland Management Plan are provided on the following pages.

PUBLIC ROADSIDES

A Vegetation Management Plan (VMP) can be thought of as quality management for roadsides. It is a decision-making system that considers a variety of tools to manage vegetation in an economically and environmentally sound manner.

Vegetation along roadsides is managed for a number of reasons including maintaining visibility for drivers, reducing water on the roadway, protecting longevity of the road surface, and minimizing fire danger.

Crews map and monitor weed treatment progress in the San Gabriel River.

VMP relies on consideration of all methods for controlling vegetation in a particular geographic location. Economics, politics, and community interest are also criteria that will influence the VMP. Mechanical management (i.e. rotary mowers, flails, reels, graders, tillers and sickles), manual management (i.e., chain saws, weed-eaters, string trimmers, shovels, scythes, hoes, and other hand-held tools), structural (hardscapes, mulches, plant selection and location) and herbicides are all appropriate tools depending on the site, funds available and management goals. Costs can be managed more effectively when they are prioritized within a VMP. Prioritization ensures that the most important activities happen first.

Goals of a Vegetation Management Program for Roadsides

Provide safe, reliable transportation corridors:
- Maintain visibility of intersections, traffic signals, curves, safety devices, signs, and railroad crossings.
- Reduce water on roadway to prevent hydroplaning.
- Keep sight lines open at intersections, driveways pedestrian crossings, known animal crossings and on curves.
- Prevent or control growth of trees and shrubs that obstruct road use.
- Minimize fire hazards.
- Prevent shading of pavement that allows formation/perpetuation of frost or ice on roadway.

Maintain the public's investment in infrastructure:
- keep ditches and other drainage structures open to prevent water from impacting road users
- Ensure drainage of water from sub base.
- Prevent pavement break-up by plants.
- Operate within budget limitations:
- Utilize all resources efficiently and economically.
- Engage local communities as partners in the VMP (abatement districts, tax assessment districts and/or voluntary programs)
- Measure actual vs. planned performance.
- Implement corrective action as needed.

Protect worker safety:
- Reduce worker risk and injury.

Minimize environmental impacts:
- Reduce the spread of weeds and control noxious weeds.
- Control erosion.
- Maintain roadway aesthetics.

Typical Methods of Vegetation Control for Roadsides

Physical methods:
    a. Hand pulling and hoeing
    b. Flaming
    c. Steaming/application of hot water
    d. Tillage (cultivation)
    e. Mowing and shredding
    f. Mulches and solarization
    g. Structural
    h. Weed mats

Chemical control:
  a. Selective vs. non-selective
  b. Pre-emergence vs. post-emergence
  c. Contact vs. systemic

With ever diminishing budgets, cost-effectiveness becomes an important criterion in the evaluation of control methods. Mowing and herbicide use have historically been the most cost-effective method for vegetation management along roadsides when compared with other methods. However, criteria other than cost-effectiveness may influence method selection in particular locations.

The Topanga Canyon Boulevard Vegetation Management Implementation Plan 2013-2017 provides an example of a plan created through a collaborative process between agencies and community groups to manage roadside vegetation through a protected area. The document includes a Best management Practices Matrix and adaptive management process. *http://www.dot.ca.gov/dist07/sync/cpimages/file/Topanga%20Canyon%20Boulevard%20Vegetation%20Management%20Implementation%20Plan%202013-2017%20v10-21-13.pdf*

WILDLANDS
(modified from The Nature Conservancy)

Wildlands present a number of unique challenges for vegetation management and may be managed for a number of different goals. Vegetation management goals in wildlands may include preservation of a specific species, maintenance of wildlife habitat, protection of sensitive resources such as streams or lakes, and providing visitor use or recreation just to name a few. In addition, wildlands often abut more developed areas, which may result in vegetation management goals mandated by law such as reduction of fire fuel loads. Wildlands present a number of unique challenges for invasive plant management. First, areas may be difficult to access, reducing the use of large equipment and the number of visits made to a site. Second, effects of invasive plant management techniques on desirable native plant, animal species, and water quality must be considered.

Considering the wide range of invasive plant problems that may be encountered in a wildland setting, instead of providing a recipe for all exotic plant problems, TNC provides a decision-making flow chart (*http://tncweeds.ucdavis.edu*) and refers to the overview chapter for information on specific techniques. Finally, check the TNC web site for a list of Internet and published resources relevant to exotic plant management in wildlands.

1. Decide upon conservation and management goals for the area being managed. Is the area primarily for wildlife habitat (e.g., a bird refuge or game reserve)? Is it focused on conservation of a single threatened plant or animal? Whatever your conservation goal, exotic plant problems should be considered in light of this goal. With the large numbers of invasive plants in California's wildlands, elimination of all non-native plants from your area may not be a realistic goal. Instead focus on species as they impact your management or conservation goal.

2. Identify invasive plant species and populations that negatively impact your goal(s). Prioritize these species and populations based on their perceived impact on your goal(s).

3. Identify possible control techniques available for this species in your area. Consider

possible negative side effects of control techniques on goals and other native plant and animal species in the area. Is consultation with US Fish and Wildlife Service and/or California Department of Fish and Wildlife necessary? It may be better to anticipate concerns of resource agencies by performing a biological survey.

4. Based on available information, develop a weed management plan that you will be able to implement with your available resources.

5. Monitor your exotic plant populations to assess the effectiveness of your management techniques.

6. Review and modify your management techniques based on monitoring information.

General Considerations When Working In a Wildland Setting

1. *Disturbance.* Most exotic invasive species rely on disturbance for establishment or thrive under disturbed conditions. Consider whether the management technique you are thinking of using will create more disturbance and potentially exacerbate your weed problem. Actions such as disking or other forms of ground disturbance need to be monitored for the potential for reestablishment and often require retreatment on a regular basis or planting of natives in the disturbed area in order to suppress exotics.

2. *Effects of pesticides on native flora and fauna.* Although pesticide applications can be very effective in wildlands for elimination of invasive species, due consideration needs to be given to the potential for pesticides to affect non-target plants or animals in the area. Consult the pesticide label and summaries of pesticide information in the TNC Weed Handbook prior to applying any pesticide in a wildland setting.

3. *Wildlife Issues.* Many species of native wildlife have come to rely on exotic plants for nesting and/or food sources. For example, the federally endangered willow flycatcher nests in both *Arundo donax* and *Foeniculum vulgare*. Prior to removing an invasive species infestation, make observations and consider possible wildlife use. If you find that wildlife are relying on this species for food or shelter, consider a phased elimination where plants are gradually removed and replaced with native vegetation that can fill a similar role for wildlife. When undertaking a phased elimination, remember that natives you plant may not be at a growth stage for wildlife use for several years. In addition, invasives that remain on site can act as seed source and spread the infestation. Sometimes this can be avoided by maintenance cutting to prevent reproduction.

Appendix A. How Pesticides Are Regulated Within California

Herbicide usage was the most controversial topic covered by the LA WMA BMP committee in developing this document; therefore, it is appropriate that additional information be provided regarding this topic to reflect the discussions and research of the committee.

REGULATORY OVERSIGHT

In the United States, all herbicides and inert ingredients must be registered for use by the US Environmental Protection Agency (EPA). EPA requires data from numerous tests on an herbicide's toxicity to people and non-target organisms and on its environmental risks before it is registered. The tests are quite extensive and take many years to perform. The State of California Environmental Protection Agency (CalEPA) Department of Pesticide Regulation may require further testing and must approve all herbicides for use in California.

ASSESSING THE HEALTH RISK OF PESTICIDES

The mission of CalEPA's Department of Pesticide Regulation (DPR) is to ensure that people and the environment are protected from adverse (harmful) effects that may be associated with pesticide use (*http://www.cdpr.ca.gov/docs/pressrls/dprguide.htm*). Determining what those impacts might be and under what circumstances they can occur is essential to an effective regulatory program. When this information is known, measures can be taken to limit exposures so that adverse effects can be avoided.

There are more than 865 active ingredients registered as pesticides, which are formulated into thousands of pesticide products available in the marketplace. About 350 pesticides are used on the foods we eat and to protect our homes and pets.

DPR scientifically evaluates the hazards of pesticides before they can be sold in California. Chemicals already in use are also subject to periodic reevaluation. Risk assessment plays a critical role in this process and is often the driving force behind new regulations and other use restrictions. DPR takes a multimedia approach to risk assessment and assesses potential dietary, workplace, residential, and ambient air exposures.

What is risk assessment?

Toxicity is an inherent property of all substances. All chemical substances can produce adverse health effects at some level of exposure. In this context, risk is the likelihood that an adverse health effect will result from an exposure (or exposures) to a particular amount (dose) of a chemical. Therefore, risk is a function of both toxicity and exposure. Risk assessment is a process designed to answer questions about how toxic a chemical is, what exposure results from its various uses, what is the probability that use will cause harm, and how to characterize that risk.

A 1997 evaluation of CalEPA risk assessment policies and practices said that although "risk assessment is known to have considerable uncertainty, and there are difficulties in applying this imperfect process to decision-making, … (it) helps prevent arbitrary decisions by providing a systematic means of incorporating scientific information into decision-making." In this light, DPR conducts health risk assessments on pesticide active ingredients to find out if they are being used (or can be used under modified conditions) in a way that is safe for both users and the general population.

The 1997 review concluded that DPR's risk assessment practices are generally consistent with the systematic scientific framework used by the USEPA and similar regulatory agencies. Where differences exist, they mostly arise from differences in law, or from situations where California differs significantly from the average for the U.S., such as in diet, climate, agricultural practices, or population demographics.

How are risk assessments conducted?

DPR, like U.S. EPA and other agencies, views risk assessment as consisting of four elements:
- Hazard identification
- Dose-response assessment
- Exposure assessment
- Risk characterization

Hazard identification involves the review and evaluation of a chemical's toxic properties – the extent and type of adverse health effects. Laboratory studies on animals are generally used to define the types of toxic effects caused by a chemical and the exposure levels (doses) at which these effects may be seen. In evaluating chemicals, scientists must determine the exposure level at which adverse effects would not be expected to occur.

Dose-response assessment considers the toxic properties of a chemical and determines the lowest dose of the chemical that results in an adverse effect. State and federal tests require that laboratory animals receive high enough doses to produce toxic effects.

Animals receive a wide range of exposures, including doses that may be much higher than those to which people might be exposed. There also are doses at which no ill effects occur in the test animals. Within that range of doses, the highest tested dose that does not cause adverse effects is the "no observed effect level" (NOEL).

Uncertainty factors are mathematical adjustments used when scientists have some but not all information. One way they are used in risk assessments is to compensate for uncertainties in the process that estimates the dose level in humans at which there is reasonable certainty that the identified adverse effects will not occur. As a default, if the toxicity studies are based on animals, we generally use an uncertainty factor of 10 to account for assumed differences in sensitivity between humans and experimental animals to a chemical (an assumption that the least sensitive humans are 10 times more sensitive than the most sensitive animal species). An additional uncertainty factor of 10 is used to address differences in sensitivity among humans (this assumes that the most sensitive human is 10 times more sensitive than the least sensitive human. This results in a total uncertainty factor of 100.

Exposure assessment is the process of finding out how people come into contact with the pesticide, how often and for how long they are in contact with the substance, and how much of the substance they are in contact with. It includes an estimate of people's potential exposure to a chemical at work, at home, or in their diets.

Exposure may be short duration (acute, occurring once or for a short time), intermediate duration (subchronic, generally one to three months), or long-term (chronic, generally one year to lifetime).

Rates of exposure are determined for breathing (inhalation), eating or drinking (ingestion), or contact with the skin (dermal absorption), depending on the chemical and the ways people may be exposed to it.

Risk characterization quantifies the results of the risk assessment (generally based on animal studies) with exposure assessment (based on estimated human exposure).

For example, characterizing the risk to pesticide applicators requires estimating what dose of the chemical caused what effects (that is, the dose response assessment), and what dose workers are exposed to (the exposure assessment). The results are often expressed in one of two ways. The first is as a margin of exposure, which is calculated by dividing the NOEL by the estimated human exposure. If the NOEL is based on a study using experimental animals, the benchmark margin of exposure would be 100 to assure that there is reasonable certainty that the effect will not occur in exposed people.

For cancer effects, risk is often expressed another way, as how much more likely it is that cancer will result from exposure to a chemical. Often, this is simplified in a kind of scientific shorthand, for example, a cancer risk of "one in a million" in a given population. This can give the inaccurate impression that science can determine that exactly one person in a million will develop cancer, that we can determine and measure the causes of all cancers. The inherent uncertainty in risk assessment means that risk assessors can only predict the probability of risk.

How does DPR collect the information used to assess risk?

DPR evaluates and registers pesticides before they are sold or used in California. The statutory guidelines require companies who wish to sell pesticides in California to submit tests and studies to DPR for evaluation. DPR's requirements for this data are very similar to those of U.S. EPA, although DPR sometimes requires some additional specific data (for example, on worker exposure, or potential to contaminate ground water). Registrants may conduct the studies themselves or hire laboratories to do testing.

Pesticide registration data requirements provide scientists with an extensive repository of information from which to make evaluations and draw conclusions. (This is not required for any other class of industrial chemicals; only pharmaceuticals are this extensively studied before use is allowed). DPR scientists also research the entire scientific literature to locate additional information on pesticides, to ensure that their conclusions are based on the most accurate, timely information on potential hazards to human health.

Do other scientists review DPR's risk assessments?

Yes, DPR's risk assessments are subject to rigorous peer review by objective, nongovernmental scientists with expertise in the scientific disciplines covered in the assessment. DPR presents the four components of the risk assessment in a risk characterization document (RCD). The RCDs also contain a risk appraisal section, which delineates the limitations, assumptions, and uncertainties in the risk assessment. The initial RCD draft undergoes internal departmental peer review by DPR scientist. After completing department review, the RCD currently undergoes peer review by scientists at the Office of Environmental Health Hazard assessment (OEHHA), another branch of CalEPA, and by scientists at U.S. EPA. DPR also uses other scientific experts for additional external peer review (e.g., scientists from the University of California). External peer review provides critical information for DPR on the scientific completeness of its documents. DPR reviews the comments, responds to the reviews, and makes changes as appropriate. In addition, as new data become available, DPR may update the RCD with appendices.

How does DPR use the results of a risk assessment?

DPR management reviews the results of the risk assessment and determines if the calculated risks are unacceptable (that is, an inadequate margin of exposure or a significant cancer risk). If risks are unacceptable, DPR then determines if risks can be controlled or mitigated. This is part of the risk management process.

What is risk management?

Risk management is the evaluation and selection of mitigation options. Risk managers use risk assessment as an important tool to determine the acceptability of a level of exposure and then reduce exposures to that level. Unlike risk assessment, risk management is not based solely on scientific considerations, since it also involves social, economic, and legal considerations to make regulatory and policy decisions. DPR considers these factors in analyzing the possible regulatory responses to potential health hazards. The process is necessarily subjective in that it requires value judgments on the acceptability of risks and the reasonableness of control measures. However, the bottom line is simple: DPR will not allow a chemical to be used unless it can be used safely.

The process of risk assessment is separate from risk management. Risk assessment often drives risk management, but risk management cannot and does not drive risk assessment at DPR. Risk assessments and risk management options are developed by separate DPR branches and are described in separate formal documents.

The National Academy of Sciences (NAS) seminal 1983 report, Risk Assessment in the Federal Government: Managing the Process, formed the foundation for the risk assessment process in general and for regulatory agencies in particular. In this report, NAS specifically addressed the separation of risk assessment and risk management. Contrary to oft-repeated misinterpretations, the report did not recommend an organizational separation of risk assessment and risk management (that is, placing the two processes in separate organizations). Rather, the report recommended the maintenance of a "clear conceptual distinction between assessment of risks and consideration of risk management alternatives; that is the scientific findings and policy judgments embodied in risk assessments should be explicitly distinguished from the political, economic, and technical considerations that influence the design and choice of regulatory strategies."

What other departments conduct risk assessment and risk management activities?

DPR is not the only State agency that conducts both risk assessment and risk management activities. The Department of Toxic Substances Control assesses exposure to various hazardous chemicals and manages the associated risks. The Department of Fish and Game assesses ecological toxicology and exposure of aquatic and terrestrial organism to various chemicals, and jointly manages the associated risks with the State and Regional Water Boards; and the Department of Health Services determines human exposure to chemicals in drinking water and

FIGURE 32. A State-licensed qualified applicator performs spot treatments along a Caltrans right-of-way. (Photo courtesy of Los Angeles County Agricultural Commissioner's Office.)

manages the associated risks. OEHHA conducts risk assessments and has a statutorily mandated "joint and mutual responsibility" with DPR for the development of regulations regarding pesticides and worker safety. The development of regulations relating to worker safety is a risk management activity.

What is the reputation of DPR's risk assessment activities?

DPR's current risk assessment activities are state of the art and widely recognized to be world class and scientifically sound. DPR separates its risk management activities from its risk assessment function, so that risk management decisions are made transparently, using the recommendations from the risk assessors. Additionally, academic experts both within and outside of California subject risk assessments to rigorous peer review.

Professional Applications Countywide Reported to the Los Angeles County Agricultural Commissioner in 2000

State Licensed Pest Control Advisors or the County Agricultural Commission recommends all professionally applied herbicide spraying *(Figure 32)*. The spraying is then reported to the County Agricultural Commissioner per established procedures. Pest Control Advisors complete State-mandated course work (continuing education) to maintain their State License. Unfortunately, only 1/3 of all pesticides applied in Los Angeles County are done so by people licensed by the State to do so. This "professional application" is performed by those agencies mandated to do so due to the nature of their work, typically government agencies and businesses. This means that 2/3 of the pesticides used in Los Angeles County are not reported to the County Agricultural Commissioner. These non-reported pesticide uses may not be applied in appropriate amounts or by appropriate procedures. Therefore, the LA WMA is committed to the education of the individuals and organizations not currently regulated to ensure that if they use herbicides they do so properly.

Per California Codes, Health and Safety Code Section 105200-105225, Doctors are required to report any medical problems related to pesticides. In Los Angeles County Doctors are required to report this information to the Department of Health Services. In fiscal year 2002/2003, 260 medical cases related to pesticides were reported of which only 14 were related to herbicides. The other 246 medical cases were related to insecticides, rodenticides, molluscicides, disinfectants, or anti-microbial products. Considering that almost 10 million people reside in Los Angeles County and many others live in neighboring Counties and work in Los Angeles County, this is a strong indication that herbicides are used appropriately and safely within Los Angeles County. Two cases of intentional ingestion of Round-up were reported in Los Angeles County in 2002. The first incident occurred on April, 2002 when Round-up was taken with a type of prescribed pain killer. A second incident was in October. Both cases were not fatal. Due to privacy issues, more information cannot be disclosed.

Appendix B. Summary of Regulations and Resource Agencies

Here we provide a summary of some of the regulations that may affect the design of a vegetation management plan. This list is NOT comprehensive. The list is divided into federal then state regulations. The California Association of Resource Conservation Districts has developed a more detailed guide to watershed project permitting that can prove useful for a vegetation management project. Download the Guide from http://www.carcd.org/docs/publications/guidetowatershedpermitting.pdf, or order a printed copy by contacting the CARCD at (916) 457-7904 or by emailing staff1@carcd.org.

FEDERAL REGULATIONS

Endangered Species Act: If a federally endangered species may inhabit the management area, this act applies. It may affect choice of control methods and timing. Early coordination with the responsible agencies can help a project go smoothly. Contact the local U.S. Fish and Wildlife Office for more information, or see http://endangered.fws.gov for more information. If endangered anadromous fish (salmon or trout) are present in the project area, contact the Fisheries section of the National Oceanic and Atmospheric Administration (formerly the National Marine Fisheries Service) at http://www.nmfs.noaa.gov/pr/species/esa/listed.htm.

National Environmental Policy Act (NEPA): If a project will involve a federal agency or permit from a federal agency (e.g., Army Corps of Engineers) then the project will fall under the NEPA. For information on this act and assistance with compliance, see http://www.epa.gov/compliance/nepa/.

Clean Water Act (CWA): This is a federal law that regulates activities that may affect water quality. If the project includes performing activities within a streambed or other watercourse, check http://www.epa.gov/region9/water/ to find out if the CWA applies. In California, State and Regional Water Quality Control Boards regulate most sections of the CWA. Locate the nearest Regional Board, and find out more about federal and state water quality protection laws at http://www.swrcb.ca.gov/. In riparian and wetland areas, some vegetation management programs may consider leaving biomass in the streambed or wetland, or may require temporary construction of roads, ramps, or other equipment staging areas. In the past, these may have required a permit from the Army Corps of Engineers under the CWA. In California, most activities involving removal of invasive vegetation are covered under a regional general permit. Contact the local Army Corps regulatory office for further information at http://www.spl.usace.army.mil/.

Clean Air Act: This federal law regulates activities that may add pollutants to the air. Prescribed burns require compliance with this law. In southern California, this and state laws protecting air quality are regulated by Air Quality Management Districts. To find out more about permitting, visit AQMD at http://www.aqmd.gov/home/regulations/compliance/open-burn/prescribed-burns.

STATE REGULATIONS

California State Endangered Species Act: California has its own Endangered Species Act (CESA), which includes some species not on the federal list. In addition, the CESA covers listed plants on private property while the federal act only covers plants on federal property or when there is a federal nexus. For more information on the CESA, see *https://www.wildlife.ca.gov/Explore/Organization/HCPB*.

California Fish and Game Code: If proposed vegetation management plans may impact wildlife, consult the California Fish and Game Code to see if its regulations apply. See *https://www.wildlife.ca.gov/Conservation/Environmental-Review*.

California Environmental Quality Act (CEQA): CEQA applies to proposed "projects" requiring approval by State and local government agencies. "Projects" are activities which have the potential to have a physical impact on the environment and may include the enactment of zoning ordinances, the issuance of conditional use permits and the approval of tentative subdivision maps. Recent changes in legislation may mean that small vegetation management projects undertaken for restoration do not require CEQA review. This information is from the CEQA webpage. For more information, see *http://resources.ca.gov/ceqa/*.

California Coastal Act: The provisions of this act may apply if the project includes managing vegetation within the Coastal Zone and need government permits to proceed. Check *http://www.coastal.ca.gov/ccatc.html* for details.

State laws regarding streambed alteration: State codes regarding streambed alteration include Fish and Game Code Sections 1600-1616. For more information, see *https://www.wildlife.ca.gov/Conservation/LSA*.

Pesticide use and licensing: The California Department of Pesticide Regulation (DPR), a state agency under the Secretary of California Environmental Protection Agency (CalEPA), regulates all pesticide use, including herbicides, in California. DPR adopted regulations to protect ground water from contamination resulting from the use of agricultural pesticides that are on the Ground Water Protection List (3CCR section 6800[a]) in specified vulnerable areas throughout California. For more information see *http://www.cdpr.ca.gov/docs/legbills/calcode/040101.htm*. Most non-home use is considered commercial and requires particular licensing and is subject to civil and criminal enforcement. Local enforcement of pesticide regulations is by the County Agricultural Commissioner. The DPR website for licensing is: *http://www.cdpr.ca.gov/docs/license/lictypes.htm*.

LOCAL REGULATIONS

There may be local laws and regulations that also impact your activities, such as local noise ordinances. Contact the local Agricultural Commissioner to find out more. *www.cdfa.ca.gov/exec/county/countymap/*

Appendix C. Summary of Regulations for Organic Pesticides

Organic herbicide usage was and continues to be controversial. Here we discuss some of the regulations and programs that bear on organic pesticides.

Organic farmers are already using a number of methods to control pests including: mechanical (careful crop selection of disease-resistant varieties), cultural (insect traps), biological controls (predator insects and beneficial microorganisms), nutrient and water management, and chemicals (rotenone and pyrethrin). These are non-synthetic pesticides, derived from natural sources not synthetically manufactured. However, natural does not mean non-toxic.

National Organic Program

The National Organic Program (NOP) seeks to ensure the integrity of "USDA organic products in the U.S. and throughout the world." The NOP Regulations includes organic Standards Certification & Accreditation, and Compliance and Enforcement. Organic Regulations are detailed in the Code of Federal Regulations in Title 7 – Agriculture Subtitle B Chapter 1 Subchapter M – Organic Foods Production Act Provisions, Part 205, National Organic Program, § 205.1 to 205.690.

According to the NOP, organic "indicates that the food or other agricultural product has been produced through approved methods that integrate cultural, biological, and mechanical practices that foster cycling of resources, promote ecological balance, and conserve biodiversity. Synthetic fertilizers, sewage sludge, irradiation, and genetic engineering may not be used." Certified growers are eligible to display the USDA Organic label on their products.

The NOP Subpart G, the National List of Allowed and Prohibited Substances (the National List) details the evaluation criteria, synthetic, nonsynthetic, and nonorganic substances that can be used in organic crop and livestock production. It also lists the substances that may be used in or on processed organic products. In general, synthetic substances are prohibited unless specifically allowed and non-synthetic substances are allowed unless specifically prohibited.

The National Organic Standards Board (NOSB), a Federal Advisory Committee, regularly meets to review all petitions and make formal recommendations to USDA. The NOSB comprises representatives from the organic industry. Any individual or organization can petition the NOSB to allow or disallow a substance or amend standards to the National List.

OMRI

OMRI, the Organic Materials Review Institute, a 501(c)(3) nonprofit, reviews pesticides and other input products against the USDA National Organic Program Rule §205.200 – 205.290, Organic Production and Handling Requirements. Those products that OMRI determines are allowed for organic use are listed in the OMRI Products List© and eligible to display the OMRI Listed® seal.

OMRI's staff and independent Review Panels rigorously assess products and materials seeking certification. Technical experts make up an Advisory Council that oversees the development of policies and standards. OMRI's Board of Directors is responsible for final approval.

Since OMRI program participation is voluntary, products not assessed or listed by OMRI may still meet USDA NOP rules. A product's absence from the OMRI Lists does not necessarily mean it does not comply with NOP standards.

OMRI publishes several lists including:
- *OMRI Products List, a Directory of Products for Organic Use*. All products are divided into sections by Category, Company, and Product Name. They are further subdivided into Crops, Livestock, and Processing then alphabetized. Products listed in the *OMRI Products List* are designated as either Allowed or Allowed with Restrictions for use under the NOP Rule.

- *OMRI Generic Materials List, a Directory of Substances Allowed and Prohibited in Organic Production and Handling*. All materials are divided into three sections: Crop Production Material, Livestock Production Materials, and Processing and Handling Materials. Within in each section materials are listed alphabetically then designated with its OMRI status: Allowed, Prohibited, or Allowed with Restrictions by the NOP Rule.

Allowed with Restrictions indicates use specific conditions that are required to be compliant with the NOP Rule.

The *OMRI Products List*, website search and free downloads are updated twice a month; the printed list is published once a year. The online *Generic Materials List* is updated as needed, and the list is printed every two years. Those seeking the most current information should check the web site *http://www.omri.org* for updates. No login or password is needed as this service is free.

The list of weed controls vary widely from plastic mulch to weak acids to mined minerals to plant extract derivatives to coconut fiber to biocontrols. The list for insect controls is equally impressive and ranges from boric acid to microbial products.

OMRI listed products can be harmful to the environment, humans, and livestock if improperly used. Therefore, all material/products should carefully follow guidelines, restrictions and regulatory annotations as well as label instructions on products.

Limitations:
OMRI lists products that are allowed in organic agricultural applications whereas we are seeking solutions to open space/wildland, water-quality, and wildlife sensitivity issues.

OMRI charges application, review and annual listing fees. If a product is not listed it may be that the manufacturer has decided against the expense of being reviewed or listed by OMRI. The product may still meet NOP standards.

OMRI does not assess synergistic effects of listed pesticides.

OMRI does not investigate the effect of listed pesticides on non-target species including endangered or sensitive species.

OMRI listing of a product does not guarantee compliance with all Federal, State and local regulations. The onus for compliance with all laws and regulation is the user's responsibility.

Little research exists on the efficacy of organic products in non-crop situations. More research needs to be published on these uses.

Glossary

Allowed*: OMRI status of materials that may be used in organic production, processing, or handling without restrictions.

Allowed with Restrictions*: OMRI status of materials that may be used in organic production, processing, or handling only under specific conditions, with certain restrictions, or as otherwise annotated.
Algacide: a pesticide used for control of aquatic algae.

Anadromous: Fish that spend most of their adult lives in salt water, and migrate to freshwater rivers and lakes to reproduce.

Best Management Practices: A practice or combination of practices determined to be an effective and practicable (including technological, economical, and institutional considerations) means of preventing or reducing environmental impacts or more specifically pollution.

Best Management Practices (BMPs): Methods, measures or practices to prevent or reduce water pollution, including but not limited to, structural and non-structural controls, operation and maintenance procedures, and other requirements, scheduling, and distribution of activities. Usually BMPs are applied as a system of practices rather than a single practice. http://www.gfc.state.ga.us/Publications/RuralForestry/GeorgiaForestryBMPManual.pdf

Disinfestate: the activity of getting rid of vermin; the application of procedures intended to eliminate arthropods which may cause diseases or are potential vectors of infectious agents of animal diseases.

Generic material*: Common name used to describe a nonproprietary substance on the OMRI Generic Material List.

Generic Materials List, OMRI: List published as part of the OMRI Standards Manual of general categories of materials used in organic crop production, food processing, and livestock production.

Herbicide: a pesticide used for control of plant species.

Insecticides: A pesticide used for the control of insects. Some insecticides are labeled for the control of ticks, mites, spiders and other arthropods.

Integrated Pest Management Plan (IMP): A pest management plan that uses life history information and extensive monitoring to understand a pest and its potential for causing economic damage. Control is achieved through multiple approaches including prevention, cultural practices, pesticide applications, exclusion, natural enemies and host resistance. The goal is to achieve long - term suppression of target pests with minimal impact on non target organisms and the environment.

LD50 or Lethal Dose: The lethal dose of a pesticide that will kill half of a test animal population. LD 50 values are given in milligrams per kilogram (mg/kg) of test animal body weight.

Molluscicides: A pesticide used to control slugs and snails.

* OMRI Glossary

National List*: A published list of synthetic materials a;;owed and natural materials prohibited in organic production, as well as nonorganic ingredients allowed in organic processing, under the provisions of OFPA.

National Organic Program (NOP)*: The section of the USDA that regulates organic production, handling, processing, and labeling.

National Organic Standards Board*: A board established by the Secretary under 7 U.S.C. 6518 to assist in the development of standards for substances to be used in organic production and to advise the Secretary on any other aspects of the implementation of the National Organic Program.

Nonsynthetic*: not synthetic, not produced by refining or reacting petroleum molecules.

No Observed Effect Level (NOEL): The maximum dose or exposure level of a pesticide that produces no noticeable toxic effect on test animals.

Organic Foods Production Act of 1990 (OFPA)*: The US federal law that defines the term 'organic.' *OMRI: the Organic Material Review Institute, a 501(c)(3) non-profit material review organization serving the organic community and the general public.

'Organic' products*: According to the NOP Rule, in order for a processed product to be labeled as 'organic' it must contain at least 95 percent organic ingredients, excluding water and salt.

Pesticide*: 1. A substance used to control insects, fungi, rodents, weeds, or other organisms that are considered pests. 2. Any substance which alone, in chemical combination, or in any formulation with one or more substances is defined as a pesticide in the Federal Insecticide, Fungicide, and Rodenticide Act (FIFRA)(7 U.S.C. 136(u)).

Pesticide Management Zones (PMZs): A geographical area, established by the California Department of Pesticide Regulation (DPR) of approximately one square mile, which is sensitive to groundwater pollution. The goal is to prevent further contamination of groundwater in areas where pesticide contamination has occurred. http://www.cdpr.ca.gov/docs/emon/pubs/protocol/136prot.pdf

Plant extract*: A substance obtained from a plant by means of a solvent without undergoing a synthetic reaction.

Products List, OMRI*: Directory of commercial products that OMRI has determined to be suitable for use in organic production, handling, and processing including company contact information. Published annually and updated quarterly.

Prohibited*: OMRI status of materials that may not be used in organic production, processing, or handling.

Registered pesticide*: Substance that is required to be registered with the Environmental Protection Agency under the Federal Insecticide, Fungicide, and Rodenticide Act.

Rodenticides: A toxic substance used for the control of rats, mice, gophers, squirrels and other rodents.

synthetic*: A substance that is formulated or manufactured by a chemical process or by a process that chemically changes a substance extracted from naturally occurring plant, animal, or mineral sources, except that such term shall not apply to substances created by naturally occurring biological processes.

USDA: U.S Department of Agriculture

OMRI Glossary

Bibliography

Berg, N., A. Hall, F Sun, S.C. Capps, D. Walton, B. Langenbrunner, and J.D. Neelin. 2015. 21st-century precipitation changes over the Los Angeles region. Journal of Climate, 28(2): 401–421. DOI: 10.1175/JCLI-D-14-003161.1

California Department of Pesticide Regulation. 2011. *A Guide to Pesticide Regulation in California.* California Department of Pesticide Regulation publication. 148 pp. http://www.cdpr.ca.gov/docs/pressrls/dprguide/dprguide.pdf

Clayton, Matthew, and Andrew Williams, editors. 2004. *Social Justice.* Blackwell Publications, Malden, Mass. 325 pgs.

DiTomaso, J.M., G.B. Kyser et al. 2013. *Weed Control in Natural Areas in the Western United States.* Weed Research and Information Center, University of California. 544pp.

Donaldson, D. R., C.L. Elmore, B.B. Fischer, K.J. Hembree, J.L. Jordan, L.S. Jordan, A.H. Lange, R. Molinar, T.S. Prather, and R. Vargas. 2002. Tree, Vine, and Soft-Fruit Crops, pp 428-430, in *Principles of Weed Control, third edition.* California Weed Science Society, Thompson Publications, Fresno, California. 630 pgs.

Elmore, Clyde, James Stapleton, Carl Bell, and James Devay. 1997. *Soil Solarization: A Nonpesticidal Method for Controlling Diseases, Nematodes, and Weeds.* University of California, ANR publication publication 21377. 16 pgs. http://vric.ucdavis.edu/pdf/soil_solarization.pdf

Emery, D. 1988. *Seed Propagation of Native California Plants.* Santa Barbara Botanic Garden, Santa Barbara, CA.

Fennimore, Steven, and Carl Bell. 2014. *Principles of Weed Control: fourth edition.* California Weed Science Society, Kindle Edition.

Griffin, D., and K.J. Anchukaitis. 2014. How unusual is the 2012–2014 California drought? Geophysical Research Letters 41(24): 9017-9023.

Holloran, Pete, Anouk MacKenzie, Sharon Farrell, Doug Johnson. 2004. *The Weed Workers' Handbook, A Guide to Techniques for Removing Bay Area Invasive Plants.* The Watershed Project and the California Invasive Plant Council. 120 pgs. http://www.cal-ipc.org/ip/management/wwh/pdf/18601.pdf

Jordan III, William, Michael E. Gilpin, and John D. Aber. 1991. *Restoration Ecology, A Sustainable Approach.* Cambridge University Press. 342 pgs.

Los Angeles County, Department of Public Works. 2014. Low Impact Development Standards Manual. Los Angeles County, Department of Public Works publication. 496 pp. *https://dpw.lacounty.gov/ldd/lib/fp/Hydrology/Low%20Impact%20Development%20Standards%20Manual.pdf*

Los Angeles County, Department of Public Works. 2010. *Stormwater Best Management Practice Design and Maintenance Manual For Publicly Maintained Storm Drain Systems.* Los Angeles County, Department of Public Works publication. 240 pp. *http://dpw.lacounty.gov/des/design_manuals/StormwaterBMPDesignandMaintenance.pdf*

National Park Service. In prep. Invasive Plant Management Plan and Environmental Assessment for Redwood National Park and Santa Monica Mountains National Recreation Area.

O'Brien, Bart, Betsey Landis, and Ellen Mackey. 2006. *Care and Maintenance of Southern California Native Plant Gardens. Cuidado y mantenimiento de jardins de plantas natives del sur de California.* Pages/paginas: viii, 238. Metropolitan Water District of Southern California, Los Angeles, California.

O'Connor-Marer, Patrick J. 2000. *The Safe and Effective Use of Pesticides.* University of California, ANR publication 3324. 342 pgs.

Perrow, Martin R., and Anthony J. Davy. 2002. *Handbook of Ecological Restoration: Volume I Principles of Restoration.* Cambridge University Press. 432 pgs.

Pimentel, David, Rodolfo Zuniga, and Doug Morrison. 2005. Update on the environmental and economic costs associated with alien-invasive species in the United States. Ecological Economics 52 (2005) 273–288.

Radosevich, Steven, R., Jodie S. Holt, and Claudio Ghersa. 2007. *Weed Ecology, Implications for Management, Third Edition.* John Wiley & Sons. 454 pgs.

Senseman, Scott, editor. 2007. *Herbicide Handbook, Ninth Edition.* Weed Science Society of America. 458 pgs.

The Low Impact Development Center, Inc. 2010. *Low Impact Development Manual for southern California: Technical Guidance and Site Planning Strategies.* Southern California Stormwater Monitoring Coalition in cooperation with the State Water Resources Control Board. 213 pp. *https://www.casqa.org/resources/lid/socal-lid-manual*

Vencill, William K., editor. 2002. *Herbicide Handbook, Eighth Edition.* Weed Science Society of America. 462 pgs.

Bibliography for "Organic" Control

About Organic Produce. No date. *http://www.ocf.berkeley.edu/~lhom/organictext.html*. Accessed 13 November 2012.

Budhu, Deodat and Jeff Charles. 2007. Sustainable Vegetation Management Practices, Chemical and mechanical means combine to control weeds. Government Engineering. March-April 2007. p. 26-27.

OMRI. Welcome to the Organic Materials Review Institute. 2011. *https://www.omri.org/*. Accessed 6 November 2012.

OMRI. 2011. *Generic Materials List, A Directory of Substances Allowed and Prohibited in Organic Production and Handling.* Slub Design, Aptos, CA. p.128.

OMRI. 2012. *Products List, A Directory of Products for Organic Use.* Slub Design, Aptos, CA. p.128.

USDA, Agricultural Marketing Service. National Organic Program. 6 June 2012. *http://www.ams.usda.gov/AMSv1.0/nop*. Accessed 10 November 2012.

US GPO, Keeping America Informed. Electronic Code of Federal Regulations. 8 November 2012. *http://www.ecfr.gov/cgi-bin/text-idx?c=ecfr&sid=3f34f4c22f9aa8e6d9864cc2683cea02&tpl=/ecfrbrowse/Title07/7cfr205_main_02.tpl.* Accessed 9 November 2012.

Internet Resources

http://www.cal-ipc.org. Homepage of the California Invasive Plant Council.

http://www.californianativeplants.com/index.php/plants/planning_tools/planting_guide. Tree of Life Nursery. *Instructions for Planting and Care: Ten Easy Steps to Success.*

http://www.cwss.org. Homepage of the California Weed Science Society.

http://water.epa.gov/polwaste/green/

http://www.fs.usda.gov/detail/okawen/recreation/horseriding-camping/?cid=fsbdev3_053645

http://www.fs.usda.gov/Internet/FSE_DOCUMENTS/fsbdev2_026443.pdf

http://www.invasivespecies.gov. A summary of government resources with respect to invasive plants.

http://tncweeds.ucdavis.edu. The Nature Conservancy's wildland weed page.

http://ucanr.edu/sites/socalinvasives. Homepage of Southern California Invasives; the website of Carl Bell, Regional Advisor – Invasive Plants, University of California Cooperative Extension.

http://www.nps.gov/goga/natural/vegmgtpl/index.htm. Summary of the vegetation management plan for the Presidio.

http://www.ipm.ucdavis.edu. The UC Statewide IPM program with general and specific information about pest control.

http://wric.ucdavis.edu. Site for the UC Davis Weed Research and Information Center, which has information on control of several invasive plants.

http://www.dot.ca.gov/hq/LandArch/roadsidehome.htm. Caltrans Roadside Management Toolbox, a web based decision making tool provided to improve the safety and maintainability of transportation projects.

http://www.ladpw.org/wmd/watershed/LA/LAR_planting_guidelines_webversion.pdf. Los Angeles River Landscaping Guidelines, County Dept of Public Works.

http://www.umt.edu/sentinel. Missoula Open Space Vegetation Management information. Includes many vegetation management plans for open space.

http://www.cityofseattle.net/parks/Horticulture/vmp.htm. Vegetation management plans for Seattle parks and open space areas (many of which are small pockets of wildland surrounded by urban areas).

http://www.mtnvisions.com/Aurora/msmgplan.html. Ecosystem management plan for an area in Arizona. Includes specific goals for different areas.

http://www.ser.org. Society for Ecological Restoration.

http://www.sercal.org. California Society for Ecological Restoration.

HAND-PULLING:

http://hortweb.cas.psu.edu/courses/hort238/Control.html

MULCHES:

http://www.nrcs.usda.gov/feature/backyard/Mulching.html

http://ohioline.osu.edu/hyg-fact/1000/1083.html

http://www.ext.vt.edu/pubs/envirohort/426-724/426-724.html

SOLARIZATION:

http://www.uidaho.edu/ag/plantdisease/soilsol.htm

http://ag.arizona.edu/gardening/news/articles/12.8.html

http://www.gardenguides.com/TipsandTechniques/solarization.htm

FLAMING:

http://www.hort.uconn.edu/ipm/weeds/htms/flweeds.htm

http://doityourself.com/gardentools/weedcontrolflamers.htm

http://www.attra.org/attra-pub/flameweed.html